C000279516

"Leading a company successfully isn't easy. And since times are changing more rapidly than ever, your leadership skills need to adapt too. The Brain-Adaptive Leadership model explained in this book provides the understanding and the practical methods for upgrading your leadership skills for the challenges to come."
Patrick De Pauw, CEO, Social Seeder

"A timely book by Dr Dimitriadis and Prof Psychogios in this transformative moment, as the world sheds its old skin. *Neuroscience for Leaders* is a clarion call to business leaders and to the world's growing human-centred economy movement. The book offers insightful and practical approaches and frameworks necessary for leading a sustainable and meaningful socio-economic environment."
Siamak Z Salimpour, Founder, Alchemy of Work

"HR professionals worldwide are under pressure to upgrade their game by incorporating innovative methods and new ideas into their practices. At the same time, uncertainty and complexity being higher than ever have reinforced the belief that a strong focus on people is the only meaningful way forward for all companies, regardless of industry or location. Putting people at the very centre of what you do requires a better understanding of what it means to be human, how we make decisions and what drives our behaviour. Since neuroscience sheds a new light into human nature, this book is an important step towards this direction. Highly recommended."
Alexis Doukas, CEO, Optimal HR Group

PRAISE FOR THE FIRST EDITION

"With leadership remaining an ongoing challenge, this book finally sets the record straight that intelligence matters not least because it combines cognitive, emotional and intuitive dimensions."
Elena Antonacopoulou, Professor, GNOSIS, University of Liverpool Management School

"I firmly believe that business results are the direct outcome of collective efforts, guided by a strong vision and even stronger leadership. However, until now the true nature of leadership was difficult to explain, regardless experiencing it daily. Dr Dimitriadis and Dr Psychogios have managed to untangle the mystery of effective leadership by revealing how the brain works in complex business environments. Read it, practise it and become a great leader!"
Maria Anargyrou-Nikolic, General Manager, Coca-Cola HBC Slovenia, Croatia, Bosnia and Herzegovina

"There are many books on leadership – perhaps too many! However, this book is well worth reading. The authors have done a really great job in bringing together developments in neuroscience and thinking about leadership to produce a thought provoking view of leadership. They make the complex area of neuroscience very accessible and use this to produce a really helpful framework for thinking about leadership. The authors have achieved a fine balance between the rigour of their arguments and the practical applications of the framework that they have developed. Even if you have a lot of books on leadership on your shelves you should make room for this one."
Malcolm Higgs, Professor of Organization Behaviour and HRM, Southampton Business School, University of Southampton

"Neuroscience is invading every scientific, economic and social aspect more and more. This book is the brilliant confirmation that understanding the mechanisms, processes and functions of the brain can help us solve problems, distinguish leaders from regular managers and uncover the right answers for generating growth."
Matteo Venerucci, Cognitive Psychologist and Brain Propaganda Founder

"An amazing compendium that will certainly assist managers and professionals in advancing their leadership skills. The book deals with the subject of this interdisciplinary research to resolve complex problems while facing challenging environments. The authors have included numerous examples from a practical and holistic perspective in order to illustrate how to create common ground about our adaptive leadership and how to put into action our leadership brain capabilities by applying their proposed BAL model. It is a fascinating book which will change your life."

Panayiotis H Ketikidis BSc MSc PhD, Professor, University of Sheffield International Faculty, City College and Chairman of the South East European Research Centre

"Leadership as an attitude? Yes. Read through and you will find it in your brain! This book will help you get acquainted with how neuroscience, psychology and anthropology can improve performance and develop your leadership brain. I have found the concepts discussed in the book highly beneficial."

Ghaleb A. Al-Ghoutani, Head of OD, Rawabi Holding – KSA

Neuroscience for Leaders

*Practical insights to successfully lead
people and organizations*

SECOND EDITION

*Nikolaos Dimitriadis
Alexandros Psychogios*

KoganPage

First published in Great Britain and the United States in 2016 by Kogan Page Limited
Second edition 2021

2nd Floor, 45 Gee Street	122 W 27th St, 10th Floor	4737/23 Ansari Road
London	New York, NY 10001	Daryaganj
EC1V 3RS	USA	New Delhi 110002
United Kingdom		India
www.koganpage.com		

© Nikolaos Dimitriadis and Alexandros Psychogios 2016, 2021

The right of Nikolaos Dimitriadis and Alexandros Psychogios to be identified as the authors of this work has been asserted by them in accordance with the Copyright, Designs and Patents Act 1988.

ISBNs
Hardback   9781789662160
Paperback   9781789662146
Ebook   9781789662153

British Library Cataloguing-in-Publication Data

A CIP record for this book is available from the British Library.

Library of Congress Cataloging-in-Publication Data

Library of Congress Cataloging-in-Publication Data
Names: Dimitriadis, Nikolaos, author. | Psychogios, Alexandros, author.
Title: Neuroscience for leaders : practical insights to successfully lead people and organizations / Nikolaos Dimitriadis and Alexandros Psychogios.
Description: Second edition. | New York : Kogan Page, 2020. | Includes bibliographical references and index. | Summary: "To be an effective leader, we need to make use of our most valued asset – our brain. As our understanding of neuroscience increases, we learn how scientific insights can be applied to develop and enhance leadership. Neuroscience for Leaders captures the most up-to-date and important findings in neuroscience and links these to the business world. This guide offers a simple framework to put these principles into practice to make better decisions, take the right actions and find faster solutions. Now in its second edition, the book presents a comprehensive approach to leading people and organizations. The authors' 'Brain-Adaptive Leadership' approach offers a step-by-step guide to enhancing the way leaders think, understanding and nurturing emotions, shaping automated brain responses and developing dynamic relationships. Examples, activities and practical suggestions are all designed to be clear and engaging. Neuroscience for Leaders is the essential guide for leaders who are ready to gain the business advantage scientifically"– Provided by publisher.
Identifiers: LCCN 2020034134 | ISBN 9781789662146 (paperback) | ISBN 9781789662160 (hardback) | ISBN 9781789662153 (epub)
Subjects: LCSH: Leadership–Psychological aspects. | Cognitive neuroscience. | Industrial management.
Classification: LCC BF637.L4 D468 2020 | DDC 658.4/092019–dc23
LC record available at https://lccn.loc.gov/2020034134

Typeset by Hong Kong FIVE Workshop, Hong Kong
Print production managed by Jellyfish
Printed and bound by 4edge Limited, UK

CONTENTS

ACKNOWLEDGEMENTS

We would like to thank the dedicated, experienced and professional team at Kogan Page for its invaluable support.

We also want to thank all our students, colleagues, corporate clients and audiences for their enthusiastic reception of the ideas in this book and for their constant provision of fresh challenges, perspectives and lessons. Our brains keep flourishing because of you!

We would like to extend our gratitude to the thousands of people around the world that bought the first edition of the book and embraced its concepts, arguments and methods. A special thank you goes to those who shared their excitement, stories and suggestions with us.

Finally, we are grateful to our families for their unlimited understanding, support and love.

The second edition of book is dedicated to everyone who desires to become a better leader through the immense, and often hidden, powers of the human brain!

ABOUT THE AUTHORS

NIKOLAOS DIMITRIADIS

Dr Nikolaos Dimitriadis is an award-winning communications professional, educator and consultant. He is the co-author of the books *Neuroscience for Leaders: A brain-adaptive leadership approach* and *Advanced Marketing Management: Principles, skills and tools* (both published by Kogan Page). He spoke at TEDx University of Strathclyde on the urgent need for Brain-Based Communication, he is a certified NeuroMarketer, and he lectured at the NeuroMarketing Manager Programme at Hamburg Media School. He is a highly sought-
after speaker and consultant, teaching regularly in MBA and corporate programmes around the world. Dr Dimitriadis has studied more than 5,000 brains in 25 countries for marketing, leadership and HR purposes. He has worked with major brands such as IKEA, IBM, JTI, Nestle, Johnson & Johnson, AstraZeneca, T-Mobile, Dell, Pierre Fabre, Coca-Cola, Banca Intesa Sanpaolo, Microsoft, CISCO, SAP, Unicredit, VMware, Emirates NBD, Raiffeisen Bank, Société Générale, Credit Agricole USAID and others. He is the Head of Neuro Consulting Services at OptimalHRGroup, offering neuro-research, neuro-advisory and neuro-training services. In 2017 he co-founded and managed the applied neuroscience start-up Trizma Neuro. He is also the Country Manager for Serbia for the University of Sheffield International Faculty, City College.

ALEXANDROS PSYCHOGIOS

Prof Alexandros Psychogios is a Professor of International HRM in Birmingham City Business School at Birmingham City University. He is also a Visiting Professor at Cyprus International Institute of Management (CIIM) and a Research Associate in South Eastern European Research Centre (SEERC). His specialization and research interests are on international human resource management; leadership, neuroscience and complexity, and performance management. His work focuses on various management and leadership issues in different but rather challenging contexts such as small and medium companies, emerging and crisis economies, project-based organizations, etc. Prof Psychogios has a wide range of experience of participating in various consultancy, training and research projects related to leadership and performance management. In August 2020, he received from the Management Consulting Division of the Academy of Management the Benedictine University Award for Outstanding Research Work on Ethical Issues in Consulting for his work on crisis leadership.

Introduction

The leadership enigma and the human brain

'What makes a great leader?' This profoundly important question has been on the minds of scientists, philosophers, professionals, and even ordinary citizens, around the globe for centuries. It is also the main question on the front cover of the November 2015 issue of the *Harvard Business Review*, a widely influential magazine for management, showcasing the obsession of business and organizations with leadership. Regardless of its importance, though, the answer to this question is perpetually elusive. Many attempts, theories and models have been proposed as an answer to the question but none has succeeded in capturing leadership in its entirety. Not anymore. We believe that the leadership enigma finally has a convincing answer. An answer rooted in a specific and very tangible source: the brain.

Leadership is not an exact science. It is not a pure art either. It is not even a set of separate thoughts, feelings or actions. Leadership is an attitude. And, as with all attitudes in classical psychology theory, it simultaneously consists of thoughts, feelings and behaviours (Rosenberg and Hivland, 1960) in a sum greater than its parts. As an attitude, leadership can be found anywhere and everywhere in social life. It is not defined by hierarchies. It is not a given qualification. We often like to view leadership as a wild flower. Wild flowers can grow in many places insofar as there are appropriate conditions. Similarly, leadership can emerge if it finds appropriate conditions within and outside organizations.

As a significant social phenomenon, leadership has been extensively studied. The main target in investigating leadership was, and still is, to identify the aforementioned conditions that influence it. Accumulative knowledge on leadership has been formulated by diverse sciences such as biology, psychology, sociology, political science, anthropology and history, among others. But, as mentioned above, regardless the amount of knowledge we currently have about leadership, we are still far from understanding the phenomenon holistically and even further from learning how to develop an effective leadership attitude. The reason is simple. Leadership is not a static phenomenon. It is a dynamic one that constantly evolves. We claim that leadership, as a dynamic phenomenon, needs a dynamic approach to understand it. In other words, we need to always expand our ability to understand leadership as an attitude based on new forms of emerging knowledge. In this direction, the real breakthrough today is neuroscience.

In the last 20 years, leaps in technology helped neuroscientists study and understand the human brain better than ever before in the entire human history. Ground-breaking insights started to emerge regarding neuroanatomy, synaptic development and brain functioning that influenced significantly the way we understand not just the brain's inner workings, but more profoundly our individual and social attitudes. As a consequence, neuroscience helped increase our understanding of leadership as an attitude within organizations. In fact, although neuroscience had a more immediate impact on other business-related fields, such as marketing and communications, there is a visible growing attention on its effects on leadership, or on what Rock and Ringleb (2009) call *neuroleadership*. These two have defined neuroleadership as the study of the biological micro-foundations of the interpersonal, influence-based relationships among leaders and their followers. Neuroleadership claims that by understanding how the human brain works, leaders can have a real advantage in engaging themselves and others towards enhanced performance (Rock and Tang, 2009). In support of this view, Henson and Rossou (2013) argued that there is already evidence that leaders become more effective when stimulating themselves and their teams to make better use of their brains' capacities. This is done mainly by enhancing individual brain strength, cultivating healthy relationships and developing high-quality collective thinking (Henson and Rossou, 2013). There are increasing voices suggesting that neuroscience opens new paths for leadership understanding and for modifying the all-important leadership attitude.

What seems to be less studied and developed, though, are practical approaches to making leaders aware of their brain flexibility and capacity, as well as how exactly those influence leadership attitudes in the real world. Although there is a relatively large number of publications on the subject, very few provide a step-by-step, concrete approach for harnessing this new knowledge in a simple, systematic and practical way for the professionals that crave it the most: business and other organizational managers.

We argue that modern leadership requires practical judgement, or what the Ancient Greek philosopher Aristotle called *phronesis*. *Phronesis* is linked to judgement resulting from context-dependent actions, as these are embedded in our daily local experiences (Stacey, 2012). *Phronesis* is a way of thinking, knowing, acting and living our daily lives (Antonacopoulou, 2012). It is acquired through practical experience and it can be understood best when we take a reflective stance towards this experience (Stacey, 2012). In turn, it helps us to formulate judgements and influence the way we behave (Antonacopoulou and Psychogios, 2015). In this respect, practical judgement is associated with our awareness of the things that can influence our judgement. Without any doubt, one of them is the brain. This book aims to help practitioners of leadership (all of us) to increase awareness around the impact of the brain on leadership behaviour. In particular, the aim of this book is to offer a practical and holistic approach to understanding and implementing the leadership brain. Drawing on the growing knowledge on the brain from neuroscience and behavioural sciences, as well as from traditional leadership and business thinking, this book uncovers achievable ways to boost your leadership brain. This is crucial for challenging business and organizational environments. By using diverse and cutting-edge global research we build a comprehensive, concrete, but at the same time simple to use approach to helping managers and professionals improving their leadership capabilities. With this in mind, we propose the brain-adaptive leadership (BAL) approach as a way of thinking, feeling and acting within organized social entities.

BAL is an attitudinal approach that individuals can follow in their attempt to consider their brain and its influence on their behaviour in order to lead projects, processes and people. This approach can be included in the wider field of applied neuroscience. The emphasis is on the 'applied', suggesting that this book is for those that need to implement new and effective leadership approaches as quickly as possible, simply because existing ones are not performing particularly well.

It is important, first, to explain in more concrete terms the three main elements on which the BAL approach is based, namely the brain, adaptive and leadership.

The brain

We live in the new century of the brain. This is what *Scientific American* declared on the cover of its March 2014 issue. For more than 20 years now, neuroscience, behavioural economics and other scientific disciplines have shattered many myths about decision-making and human behaviour, and especially the stubborn fixation on the notion that logic rules. This belief is rapidly going out of fashion, since our capacity for rational and analytical thinking is found not to be *the* absolute human capability. Important? Yes. The most important? No. Neuroscience, as we discuss in detail in this book, is clear: humans are not, and should not be, pure rational creatures. Replacing rationality with a more complex, deep and, dare we say, holistic view and understanding of the inner workings of the human brain and of how it influences our lives is crucial to improving human behaviour. Consequently, in the last two decades or so, numerous studies and books have been published on neuroscience, the brain and the hidden forces shaping behaviour. At the same time, there is a growing interest from people in various professions and fields to learn these brain-based rules of human behaviour.

Advancements in medical technology have proved to be extremely helpful in this regard. Imaging technologies (such as functional magnetic resonance imaging – fMRI) and other neural, biometric and brainwaves-capturing technologies have contributed dramatically to the way the brain is studied (Gorgiev and Dimitriadis, 2016). The key outcome is that many new insights about neural processes, development and connectivity of have been uncovered. These findings influence not just the way we understand brain function per se. More importantly, they have a profound impact on revealing the ways that brain functions influence our individual and collective decision-making and, consequently, personal and social behaviour. In other words, these developments are drastically developing our understanding of social phenomena, like leadership and change, in organizations. Let's explore in simple terms, as much as we can, what our brain is, biologically speaking.

Our brain is not one organ. Instead, it is a very complex synthesis of different parts that are also dynamically evolving through the millennia. A more concrete description of the (not only human) brain is that it consists of an enormous number of neurons that are connected to each other in millions of forms called synapses. Neuroscientists claim that humans have around 86 billion neurons, and each neuron has around 1,000 synapses (*New Scientist*,

2017). Consequently, the total number of synapses that are activated in our brain is enormous, demonstrating its complex nature. A way to imagine neuro-connections is to think of a network design, as Figure 0.1 shows.

Neurons communicate with each other through electrical signalling called 'action potentials', which carry just 0.1 volts and can travel long distances at a speed of 120 meters per second (O'Shea, 2005). This electrical signalling reaches a synapse, triggering neurotransmitters that in turn cause electrical switches on the neuron. Neurotransmitters can excite or inhibit neuron activity. In general, we know that neurotransmitters' activity is essential for our brain function since it allows the brain to continuously change and adapt to the stimuli from the external and internal worlds. All in all, a brain is a web of neurons that follow a very sophisticated form of action and a very complex organization.

FIGURE 0.1   Neuron network

Beyond the above brief description of neurons in the brain, it is important to also have a clear view of its anatomy. Since birth, out brain develops various forms that all look alike. However, in order to have a better understanding of its structure we can use the simplest way to describe it, by dividing it in three main parts, which are always interconnected to each other. This is the popular triune brain theory that will be discussed in relation to persuasion in Chapter 8. Although oversimplified and highly criticized, the triune, or three, brain theory is very effective as an educational and introductory model to the brain's structure and function.

The first brain is known as the brain stem or the hindbrain or, simply, the reptilian brain, since it can be found in reptiles, among other creatures. It is located at the base of out skull above the neck. In humans this brain has three structures: *medulla oblongata*, *pons* and *cerebellum* (Sapolsky, 2017).

The lizard brain, as it is sometimes called, is responsible for all automatic behaviours that aim to keep us alive. Functions like breathing, heartbeat, blood pressure, swallowing and sleep are linked to this part of the brain. Recent developments in brain science found that it also influences our emotional and cognitive abilities (*New Scientist*, 2017).

The second part of the brain is called the limbic system, or the emotional brain. This brain, also known as the mammalian brain, is linked to our emotional system and it is responsible for our social behaviours, including physical actions and decisions, and to basic emotions, since the amygdala (Barry *et al*, 2017), our fear and arousal centre, as well as the nucleus accumbens (Goto and Grace, 2008), our reward centre, are located there. The limbic system is linked to other human functions like hearing, visual information processing and, of course, our moods (O'Shea, 2005). One thing that needs to be clear at this point is that although this brain is linked to emotions, this does not mean that emotions 'exist' and are active only in this part of the brain. Instead we need to understand emotions as complex mental states related to the nervous system and brought about by chemical changes linked to thoughts and feelings (Panksepp, 2005).

The last part is called the neocortex or the forebrain or, simply, the rational brain. The neocortex is associated with higher processing of information and cognitive skills as well as the formation of memories. For example, the cerebral cortex is associated with our ability to put words together and speak a language as well as to be creative, to imagine things and to generate ideas (*New Scientist*, 2017). Cognitive capacities related to information processing and information storage are also key functions of the rational brain (Kaas and Herculano-Houzel, 2017). The third brain can be further structured in four principal lobes, namely the *frontal, occipital, temporal* and *parietal* lobes, where the aforementioned functions are located.

For us, it is all in the brain. The brain defines our behaviour, including emotions, thoughts and actions. Since, as we said at the start of this chapter, leadership is an attitude, we can argue that leadership is also defined by our brain's function. Leadership research and practice should start and end with the brain: a single brain as much as collective brains, within and outside organizations. Discussing leadership without neuroscience, and also psychology, anthropology and behavioural economics, is not only inadequate but can be seriously misleading. Our thought patterns, analytical skills, moods, emotional reactions, habits, relationship building and communication skills, our ability to change fast and to understand others fast, our overall influence and power of persuasion, and almost anything else you can come up with concerning leadership can be traced back into the brain. Your brain makes you a great leader. Do you allow it to do so?

Adaptive

One word can be used in order to characterize modern societies and the collective entities within them: complexity. Societies, as well as organizations, as systems of infinite human interaction, can be seen as inherently complex. We usually ask students in our lectures to define complexity in the easiest and most convenient way they can. After discussion and various definitions suggested by students, we often conclude that complexity reflects a situation, any situation that cannot be fully understood. In other words, complex is something that we cannot easily understand and ultimately control. It is important to note here that there is a critical distinction between *complicated* and *complex*. Complicated is something that depends on many factors that require, often, expert knowledge in order to explain it and place it under control. Complex is something that also depends on many factors. However, these factors are almost impossible to predict, identify and ultimately place under human control.

Complexity science emerged the last 50 years and drastically changed the way we understand physical phenomena. Complexity and chaos theory influenced social sciences too, including management and leadership, mainly during the last three decades. In this context, complexity theory considers societies and organizations as complex adaptive systems comprised of numerous autonomous agents, which engage in non-linear, unpredictable and emergent behaviour (McMillan, 2006; Psychogios and Garev, 2012). These systems have an innate capability to self-organize, since relationships within them are guided by continuous feedback loops (Kaufmann, 1993; Stacey, 2004). Similarly, chaos theory implies that social systems constantly evolve over time, as they are sensitive to small changes that can make the whole system fluctuate unpredictably (Wheatley, 1999; Cilliers, 2000).

Complexity in social sciences has become as a strong paradigm. The realization that we live in a volatile, uncertain, complex and ambiguous (VUCA) world (Horney *et al*, 2010) is widespread. In order to understand better the VUCA world we live in, just think of one day in your life. Consider for a moment how much information you receive daily, related directly or indirectly to your job. Think about your daily communication flux – how many emails you send and receive, regardless of whether they are all relevant and, even more important, useful; how many times you interact with others on a daily basis, either through face-to-face communication or through phones and digital meetings platforms such as Skype; how much time you spend reading job-related news, visiting relevant websites, etc; and how long each day you spend on work-related social media. Finally, consider the total information you receive from the wider media landscape. Now, remember that the same is happening to most individuals, in one way or another, on this planet. The resulting web of interactions is mind-blowing. This is just a small thought exercise for the world we currently live in.

The world consists of dynamism, fast-paced evolution, non-predictability, huge uncertainty and, of course, of endless information. Without any doubt, our VUCA world is an *infocratic* one. If the previous paradigm of social organization was *bureaucracy*, the new paradigm is *infocracy* (Clawson, 2011). This paradigm claims that the source of power is no longer hidden within positions and offices. Because of the information explosion we experience, it is redistributed all over our organized societal entities. Infocracy requires a new approach, oriented towards continuous adaptation and not towards a single optimum outcome, as in the bureaucratic paradigm. Infocracy claims that everything happening around us is somehow being co-created by all of us, even if some have more influence over the result (Stacey, 2010). In other words, the new paradigm elevates people as the critical component of any social system. People, and especially leaders, need to understand that they play a critical role in formulating the VUCA world around us. They are not passive actors. To do so, though, they urgently need to develop new skills of adaptation.

Brain science brings a message of hope here. This hope is called *neuroplasticity*. Our brain does not remain unchanged during our lifetime. On the contrary, it changes constantly, every single day. Neuroplasticity is the proven ability of the brain to change, be trained, adapt, grow new neural connections or degrade the existing ones. All these depend, apart from the apparent genetic factor, on us, the environment we live and on how we interact with it. Even in our adult and later years of our lives our brain constantly changes (Hood, 2014). Consequently, the brain is the key to the new kind of leadership required to continuously adapt in a complex world. This is probably why the prestigious *Harvard Business Review* devoted a whole special issue to the subject, at the start of 2019, titled 'The brain science behind business'.

Leadership

Leadership is an extremely popular term, with a significant impact on our societies and organizations. Indicative of this fact is that when you search the term 'leadership' in Google you receive approximately 4,380,000,000 results. When you search 'leadership in business' you receive approximately 1,730,000,000 results. Leadership matters. Since the beginning of human-kind and the development of primitive communities into the first civiliza-tions, leaders have played a significant role. But why do people believe so much in leadership? A critical answer can be found in psychology. According to Maccoby (2004) people are keen to follow a leader both for rational and irrational reasons. The former is about dealing with the unknown, which usually means people's future. Leaders can infuse stability and certainty when dealing with uncertainty situations. They can give hope and set a clear path in front of us. However, irrational reasons seem to be more critical. People follow a leader, often without being aware why they do so, because it is a natural response to difficult situations. This automatic willingness to follow leaders emerges from emotions and images in our unconscious mind that are projected onto our relationships with leaders (Maccoby, 2004). Leadership is an evolved state in our brains.

OUR APPROACH TO WHO IS THE LEADER

It is critical to mention that when we use the term 'leader' in this book we refer, most of the time, to all those in formal managerial positions of different levels in various types of organizations. However, it is important to remember that informal 'leaders' can also learn critical lessons from this book. It is not neces-sary to have formal hierarchical authority in order to motivate others, direct them towards a specific destination and support them during this process. Yes, we refer to all of you, mothers, fathers, friends, and many others, that you think and act like leaders without necessarily realize it. This book is also for you!

There are literally thousands of studies attempting to understand, explain and, paradoxically, somehow 'control' the phenomenon of leadership. Those exploring the literature on leadership (Pierce and Newstrom, 2014; Obolensky, 2014; Western, 2013; Stacey, 2012; Psychogios, 2007) catego-rize it in four main approaches or leadership styles:

1 The *autocratic-controlling approach*, where the leader seeks and controls resources to enhance efficiency.

2 The *motivational-engagement approach*, where the leader focuses on relationships and on motivating others in order to enhance results.

3 The *transformational approach*, where the leader aims to transform the culture of an organization, attempting to achieve better outcomes.

4 The *adaptive approach*, where the leader sees organizations as complex systems that are evolving and therefore trying to constantly adapt.

There is an evolutionary pattern above. Leadership started with control, moved to motivation, then to transformation and finally to adaptation. Our approach, as explained in the book, does not simply belong in the fourth style. It strives to expand it. Adaptive leadership deals heavily with systems, and places them at the centre of analysis (Obolensky, 2014). It is the systems that are adaptive and leaders have to operate in a constantly changing environment within those systems. In our approach, the focus is not on systems, the external world, but on the brain, the internal world. For us, the brain is adaptive, too. And it is this phenomenal capacity of the brain to change that brings leadership to the forefront of creating better organizations and a better future for all.

In particular, our approach follows a more evolutionary and biological perspective. We view leadership as rooted, among other things, in our brains and the way that they have been evolved throughout the history of humankind. A lot of arguments are based on a relatively new perspective on leadership called evolutionary leadership theory (EvoL theory) (Vugt and Ronay, 2014). EvoL theory follows a Darwinian logic and attempts to explain the way that leadership has been formulated through 'selection pressures operating on humans in ancestral environments' (Vugt, 2018, p 192). This perspective claims that human behaviour, therefore leadership, is an outcome of physiological, neurological and psychological processes influenced by biological evolution. Also, it is critical to understand that these biological products operate automatically (Kahneman, 2011), and are not context free, since they are highly sensitive to environmental influences (Tooby and Cosmides, 2005). EvoL theory uses various approaches to study leadership behaviour and outcomes. For example, organizational cognitive neuroscience (Lee *et al*, 2012), neuro-organizational behaviour (Beugré, 2018) and organizational neuroscience (Becker *et al*, 2011) apply neuroscientific tools and techniques to investigate organizational phenomena including leadership. Also, other scholars study individual difference by focusing on hormones and neurotransmitters and the way that they affect finally leadership behaviour (Van der Meij *et al*, 2016; Grouda *et al*, 2019), while others emphasize the genetic relevance of leadership roles (De Neve *et al*, 2013). Finally, some of them focused on facial expressions and their influence on leadership perceptions (Vugt and Grabo, 2015).

Beyond the above developments in EvoL theory, there are two more aspects that are extremely important in the study of the biological influences on leadership behaviour. The first is the follower's perspective. EvoL theory highlights the importance of the leader as well as the importance of the follower (Brown, 2018). As leadership is the outcome of an evolution process, the same applies to followership. In other words, followers are essential aspects in the leadership process and their input is first and foremost a product of biological evolution. The second aspect is the context perspective. EvoL theory claims that different leadership styles are a result of different circumstances occurring in different situations (Little *et al*, 2007). In similar vein, EvoL theory claims that leadership attributes are also culturally sensitive (Boyd and Richerson, 2006).

Based on the above, our approach to leadership is based on three elements: the leader, the follower and the context. We claim that leadership as a social phenomenon needs to be understood within the frame of these three elements. By separating these elements and studying them individually we can only have a limited comprehension of leadership. The common denominator of these three elements is the human brain. Focusing on the brain and understanding its influence on our behaviour we can: increase awareness of ourselves as leaders and understand better our limitations and abilities; increase awareness of others (followers) by understanding how others make decisions and are motivated; and finally increase awareness of the context by understanding special features of the time, space and situation in which leadership occurs. All in all, we strongly believe that by understanding how the brain works and by accepting our ability to alter how the brain reacts to stimuli, we are in a better position to reflect on experience and formulate judgements, and therefore to adapt our behaviour better towards ourselves and others. The BAL approach is critical in doing so.

The BAL approach in brief

The BAL approach consists of four pillars. These are the four main groups of ideas, scientific insights and practical recommendations that we have gathered, organized and used as a comprehensive brain-based approach in our business, managerial, educational and personal lives. In creating this approach, we used a large quantity of scientific and business literature focusing on the latest possible insights and combined them with classic ones. We have used academic studies, examples, case studies, personal stories, professional opinions and practical tips from around the world. But, most importantly, we have used the experience of our own consulting, research, training, coaching, entrepreneurial and managerial endeavours. We briefly describe each one in Figure 0.2.

FIGURE 0.2   The BAL approach

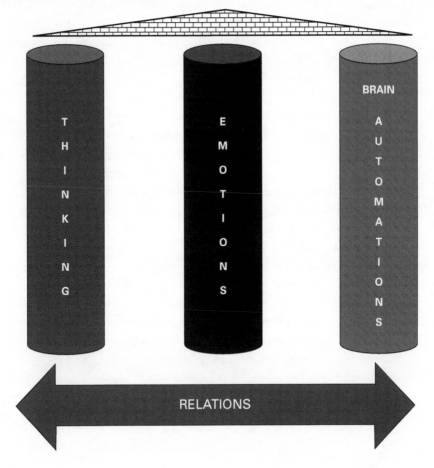

Pillar 1: Thinking

This first pillar reflects the cognitive function of our brain. It is the longest pillar in this book because leaders and businesses are in great need of learning how to think in a brain-based way. If they do so, they can release the true power of their analytical thinking and achieve more. Becoming a brain-based leader first and foremost means maintaining and increasing willpower. The why is plain and simple: leadership is about keeping your brain strong and able to deal efficiently and effectively with demanding situations. Your thoughts and decisions will never be clear, relevant, useful and inspiring if you waste your brain energy easily. Learn how to save and use your will-power, avoiding the pitfalls of ego depletion, burnout and multitasking. Awareness of cognitive biases is of equal importance. Economists, psychologists, business authors and consultants alike see biases as the main evil of non-rational thinking. A strong brain that is full of willpower but is also full of its own illusions is no good to anyone. So, leaders need to be always aware of biases in order to consciously decide when these biases are helpful (yes, they can be) and when they are not. Beware of biases and use them to your advantage when needed. Similarly, beware of automatic pattern recognition that can be misleading, and ask great questions to reveal the key elements in each situation. Questioning is a core leadership skill. Adopting a deep and meaningful personal purpose helps leaders to keep their thoughts on track. At the same time, the ability to find and get into the optimum mental state of 'flow' will help leaders boost clarity, effectiveness and productivity of thinking. Creative thinking emerges as the most important type of thinking, since resourceful problem solving is necessary in dealing with disruption and constant change. Memory often appears in surveys as the first cognitive capability people want to improve. A good working and long-term memory always impresses people around us, and will often help to win an argument. Leaders need to learn how to improve and use their memory. Along with memory, they also need to improve their mindset by adopting an open, positive, growth-focused mindset that will keep them sharp and on top of challenging situations.

Pillar 2: Emotions

The second pillar reflects the emotional life of our brains. Humans are not always able to understand and articulate their own emotions and this creates a huge challenge for leaders, taking into account that what moves us is emotions, not thoughts. We provide concrete models on naming and categorizing our core emotions and how we can recognize them in ourselves and in others. Being able to spot the right emotion early and to deal with it accordingly is key to better leadership. Also, evoking the right emotion at the right moment is integral to modern leadership. Emotional equations and emotional agility can do miracles for leaders when they are used appropriately. We explore emotional styles, the established ways we emotionally react to stimuli, and how to identify them. More importantly, we focus on how to change them to fit a modern leadership style. Leaders need also to examine their moods, or semi-stable emotional states, because a wrong mood can destroy the collaborative spirit of a team, a department or a whole organization. Find your growth, positive mood and transmit it to those around you. You will reap immediate rewards.

Pillar 3: Brain automations

The third pillar reflects the automated responses and protocols of our brains. Since our decisions and actions are primarily based on brain processes beyond our conscious control, rooted in deeper brain structures, it is imperative that modern leaders understand and use these processes to their advantage. Priming, the process by which leaders can nudge their brain and the brains of others towards a specific decision or behaviour, should be applied daily. Through priming, leaders can improve productivity, creativity and social connectivity. Similarly, habits, the most efficient way for the brain to save precious energy, should become strategic elements for enhancing leadership capacity. As patterns of repeated behaviour, habits in working environments are representative of the culture of the organization and need to be carefully approached. Recent literature suggests that changing habits and creating a set of favourable ones, both for ourselves and others, is not difficult if we follow the right process. Destroy negative habits and create positive ones to boost your leadership effect. Furthermore, our brains have evolved over millions of years to deal with the challenges and opportunities of the physical world around it. Thus, it has developed unique ways to automatically interact and respond to changes in its surroundings. In this respect, we suggest that dealing with the physical environment or physical context affects leadership decision-making and behaviour subconsciously but massively. The modern leader needs to understand the interaction between the brain and the physical world and use spaces to encourage desired behavioural responses.

Pillar 4: Relations

This fourth and last pillar reflects the social aspect of our leadership lives. We consider it the most important pillar, since it is the space where leadership as a social interaction actually takes place. This pillar should not be considered as a different brain section, but instead is the pillar that connects all the others. No leader can inspire action if others do not connect with the leader in deep and meaningful ways. And in order to connect, leaders have to fully appreciate the socially based origins of consciousness and to adopt a 'collaborate first and then reciprocate' attitude in business interactions. Dos and don'ts are provided, and modern leaders are encouraged to develop their own radar for identifying and connecting to people around them. This skill alone will dramatically increase your leadership effectiveness. Utilizing your brain's tendency for imitation in order to create and maintain an effective corporate culture, and establishing and managing extensive human networks, both through strong and weak ties, will elevate your leadership status. Furthermore, managing the art of clicking with other people by demonstrating presence, warmth and strength will help you create powerful connection with important individuals inside and outside your organization. However, if you do not make sure that relevant brain chemicals are in a collaborative mode, you will not be able to adapt your brain for networked leadership.

Relations are built and nurtured through communication. And communication for us is about creating a desired behavioural response. Thus, persuasion is a core leadership skill because modern leaders need to influence behaviours to achieve their organization's goals. In this direction, we need to 'speak' to all three main brain functions – think, feel and do – if we are to induce meaningful behavioural change. So we need to direct rationality, motivated by strong emotions, and tweak the environment for desired habits. To get maximum results both for individual and collective brains, leaders need to use, when possible, the principles of persuasion; specific phrases (like 'because') that can fast-track their influence on others, compassionate conversation and the six key stimuli of getting a brain's attention. Your increased ability to persuade others will impact positively both on your and other people's work.

Now, get ready for immersing yourself, and your brain, in the BAL approach.

Keep in mind while reading this book

1 Although neuroscience is the leading source for the BAL model, we have also used psychology, sociology, behavioural economics, management and leadership science, and even marketing and communications, in order to explain and support our approach. Today, modern leaders need all the help they can get, wherever this comes from.

2 Refer to other studies to expand on issues you find more interesting. It is impossible to expand on all the topics, concepts, examples and recommendations in a single book. However, we have provided an extensive and comprehensive list of sources, which we urge you to explore. These include those who are sceptical about the popularity of neuroscience and the topics discussed in this book, like comedian Robert Newman in his 2017 book *Neuropolis: A brain science survival guide*, psychologist Paul Bloom in *Against Empathy: The case for rational compassion* (2017), and psychologist Benjamin Hardy in *Willpower Doesn't Work: Discover the hidden keys to success* (2018). In general, close observers will notice that there is an increasing clash between classical psychology and modern neuroscience, since neuroscience with the use of brain-reading technology revises and even rejects traditional theories and methods from psychology (Dimitriadis and Dimitriadis, 2018). Having a well-informed opinion after reviewing conflicting arguments and adopting a critical approach are crucial skills for all professionals today.

3 The short stories at the start of each chapter are real. The chapters open with a leader in a difficult business situation. All those stories come from our personal work with leaders and managers throughout the world. We have changed details so they cannot be recognized personally, but the challenges remained intact.

4 All exercises and recommendations in the book have been tried and tested during the last 14 years we have spent building the BAL approach. Find the ones that work best for you and your team. To do so, experiment with as many as you can at appropriate moments. Go for quantity as well as quality.

5 This is a book for practitioners by practitioners. Not a book by neuroscientists for leaders. This is important. We do not provide details about the brain for the sake of it. We have hand-picked and carefully synthesized information we found to matter the most in our managerial,

consulting, teaching and research practices. Contrary to most books on the subject, we have not used a single brain image in this book. Leaders have no need to know about detailed brain anatomy. They have a huge need to be brain-friendly. Although we do provide specific information on crucial brain functions, we always focus on practicality. This is what the BAL approach is about.

6 The brain is still full of surprises, so keep an open mind. Neuroscience, and other fields that deal with human nature, and especially with human behaviour, are still evolving. Technology and scientific enquiry will push the boundaries of understanding, and more ground-breaking insights are about to emerge. For example, in 2018 the neuroscientist George Paxinos and his team at Neuroscience Research Australia announced the discovery of a new brain part, unknown until then, which is, according to the research team, 'absent in the rhesus monkey and other animals that we have studied. This region could be what makes humans unique besides our larger brain size' (Gander, 2018). So, never stop searching, never stop asking, and never stop learning.

7 Don't just read, do. You learn more and faster with doing than with reading. Start applying ideas and recommendations as soon as you feel like it. Experiment first in low-risk situations and slowly but surely move to more important tasks. We do not expect that you will do them all, but we do expect that you will enjoy exploring most of them.

8 Forget *Homo economicus*. This outdated view of humans, supported for centuries by some economists, is over and done. Humans are not rational, and they will never be. And that's actually great. Rationality plays a crucial role in implementing deeper decisions, but without the moral compass, the motivational boost of emotions and the steady guide of empathy we would just be cold psychopaths. Great leaders are not that. They are caring, understanding, passionate and visionary beings (with all of those characteristics or a combination of them) that move forward, no matter what. If purely rational, they would have stopped many times along the way.

9 It is about you personally and the people around you. The brain-adaptive leader changes not only one brain but many. Leaders that change their brains inevitably change the brains of those close to them, becoming an unstoppable force of progress. No leader is an island. Leaders are powerful social actors with a huge responsibility for improving everything and everyone, all the time.

10 We are all leaders. Front-line employees, middle managers and supervisors, unit and departmental directors, chief officers and chief executive officers (CEOs): all leaders. The era of non-leader professionals is over. Complexity, disruption, abundant information and the most demanding clients ever have made the old model obsolete. Whatever the position, lead as a true brain-based leader and you will bring about change that matters. A lot of it.

It is time to move on to the first pillar.

References

Antonacopoulou, EP (2012) Leadership: making waves, in *New Insights into Leadership: An international perspective*, ed H Owen, Kogan Page, London

Antonacopoulou, E and Psychogios, A (2015) Practising changing change: How middle managers take a stance towards lived experiences of change, *Annual Meeting of the Academy of Management*, August, Vancouver, Canada (Conference Proceedings – Paper 14448)

Barry, TJ, Murray, L, Fearon, P, Moutsiana, C, Johnstone, T and Halligan, SL (2017) Amygdala volume and hypothalamic–pituitary–adrenal axis reactivity to social stress, *Psychoneuroendocrinology*, **85**, pp 96–99

Becker, WJ, Cropanzano, R and Sanfey, AG (2011) Organizational neuroscience: Taking organizational theory inside the neural black box, *Journal of Management*, **37**(4), pp 933–61

Beugré, CD (2018) *The Neuroscience of Organizational Behavior*, Edward Elgar Publishing, Cheltenham

Bloom, P (2017) *Against Empathy: The case for rational compassion*, Vintage, London

Boyd, R, and Richerson, PJ (2006) Culture and the evolution of the human social instincts, *Roots of Human Sociality*, pp 453–77

Brown, JD (2018) In the minds of the followers: Follower–centric approaches to leadership, in *The Nature of Leadership*, ed J Antonakis and DD Day, pp 82–108, Sage, London

Cilliers, P (2000) Rules and complex systems, *Emergence: Complexity and organization*, **2**(3), pp 40–50

Clawson, J (2011) *Level Three Leadership*, 5th edn, Prentice Hall, London

De Neve, JE, Mikhaylov, S, Dawes, CT, Christakis, NA, and Fowler, JH (2013) Born to lead? A twin design and genetic association study of leadership role occupancy, *The Leadership Quarterly*, **24**(1), 45–60

Dimitriadis, N, Ney, J and Dimitriadis, NJ (2018) *Advanced Marketing Management: Principles, skills and tools*, Kogan Page, London

Gander, K (2018) Endorestiform nucleus: Scientist just discovered a new part of the human brain, www.newsweek.com/endorestiform-nucleus-new-part-human-brain-discovered-scientist-1228367 (archived at https://perma.cc/X4G6-6CPT)

Gorgiev, A and Dimitriadis, N (2016) Upgrading marketing research: Neuromarketing tools for understanding consumers, in *Handbook of Research on Innovations in Marketing Information Systems*, ed T Tsiakis, IGI Global (formerly Idea Group Inc), Pennsylvania

Goto, Y and Grace, AA (2008) Limbic and cortical information processing in the nucleus accumbens, *Trends in Neurosciences*, 31(11), pp 552–58

Grouda, J, Psychogios, A and Melnyk, Y (2019) The power of temperament: Exploring the biological foundations of leadership perceptions, *Interdisciplinary Perspectives on Leadership Symposium*, Theme: Power and Leadership, May, Corfu, Greece

Hardy, B (2018) *Willpower Doesn't Work: Discover the hidden keys to success*, Piatkus, London

Henson, C and Rossou, P (2013) *Brain Wise Leadership: Practical neuroscience to survive and thrive at work*, Learning Quest, Sydney

Hood, B (2014) *The Domesticated Brain*, Pelican Books, London

Horney, N, Passmore, B and O'Shea, T (2010) Leadership agility: A business imperative for a VUCA world, *People and Strategy*, 33(4), pp 34–38

Kaas, JH and Herculano-Houzel, S (2017) What makes the human brain special: Key features of brain and neocortex, in *The Physics of the Mind and Brain Disorders*, ed I Opris and MF Casanova, pp 3–22, Springer International Publishing, New York

Kahneman, D (2011) *Thinking, Fast and Slow*, Macmillan, Basingstoke

Kandel, RE (2018) *The Disordered Mind: What unusual brains tell us about ourselves*, Farrar, Straus and Giroux, New York

Kaufmann, SA (1993) *Origins of Order: Self organization and selection in evolution,* Oxford University Press, Oxford

Lee, N, Senior, C and Butler, MJ (2012) The domain of organizational cognitive neuroscience: Theoretical and empirical challenges, *Journal of Management*, 38(4), pp 921–31

Little, AC, Burriss, RP, Jones, BC and Roberts, SC (2007) Facial appearance affects voting decisions, *Evolution and Human Behavior*, 28(1), pp 18–27

Maccoby, M (2004) Why people follow the leader: The power of transference, *Harvard Business Review*, September, https://hbr.org/2004/09/why-people-follow-the-leader-the-power-of-transference (archived at https://perma.cc/85MG-7F3G)

McMillan, E (2006) *Complexity, Organizations and Change: An essential introduction*, Routledge, Abingdon

New Scientist (2017) *How Your Brain Works: Inside the most complicated object in the known universe*, John Murray Learning and *New Scientist*

Newman, R (2017) *Neuropolis: A brain science survival guide*, William Collins, London

O'Shea, M (2005) *The Brain: A very short introduction*, Oxford University Press, Oxford

Obolensky, N (2014) *Complex Adaptive Leadership: Embracing paradox and uncertainty*, 2nd edn, Gower Publishing Limited, Farnham

Panksepp, J (2005) *Affective Neuroscience: The foundations of human and animal emotions*, Oxford University Press, Oxford

Pierce, TJ and Newstrom, WJ (2014) *Leaders and the Leadership Process: Readings, self assessments and applications*, 6th edn, McGraw-Hill, New York

Psychogios, A (2007) Towards the transformational leader: Addressing women's leadership style in modern business management, *Journal of Business and Society*, 20(1 and 2), pp 169–80

Psychogios, A and Garev, S (2012) Understanding complexity leadership behaviour in SMEs: Lessons from a turbulent business environment, *Emergence: Complexity and organization*, 14(3), pp 1–22

Rock, D and Ringleb, AH (2009) Defining NeuroLeadership as a field, *NeuroLeadership Journal*, 2, pp 1–7

Rock, D and Tang, Y (2009) Neuroscience of engagement, *NeuroLeadership Journal*, 2, pp 15–22

Rosenberg, MJ and Hovland, CI (1960) Cognitive, affective, and behavioral components of attitudes, in *Attitude Organization and Change: An Analysis of consistency among attitude components*, ed MJ Rosenberg, CI Hovland, WJ McGuire, RP Abelson and JW Brehm, pp 1–14, Yale University Press, New Haven

Sapolsky, RM (2017) *Behave: The biology of humans at our best and worst*, Penguin, London

Stacey, RD (2004) *Strategic Management and Organizational Dynamics: The challenge of complexity*, 4th edn, Prentice Hall, London

Stacey, RD (2010) *Complexity and Organizational Reality: Uncertainty and the need to rethink management after the collapse of investment capitalism*, 2nd edn, Routledge, Abingdon

Stacey, RD (2012) *Tools and Techniques of Leadership and Management: Meeting the challenge of complexity*, 1st edn, Routledge, Abingdon

Tooby, J and Cosmides, L (2005) The theoretical foundations of evolutionary psychology, in *The Handbook of Evolutionary Psychology*, ed DM Buss, pp 5–67, John Wiley, Chichester

Van der Meij, L, Schaveling, J and van Vugt, M (2016) Basal testosterone, leadership and dominance: A field study and meta-analysis, *Psychoneuroendocrinology*, **72**, pp 72–79

Vugt, MV (2018) Evolutionary, biological, and neuroscience perspectives, in The Nature of Leadership, ed J Antonakis and DD Day, pp 189–217, Sage, London

Vugt, MV and Grabo, AE (2015) The many faces of leadership: An evolutionary–psychology approach, *Current Directions in Psychological Science*, **24**(6), pp 484–89

Vugt, MV and Ronay, R (2014) The evolutionary psychology of leadership: Theory, review, and roadmap, *Organizational Psychology Review*, **4**(1), pp 74–95

Western, S (2013) *Leadership: A critical text*, 2nd edn, Sage, London

Wheatley, JM (1999) *Leadership and the New Science*, 2nd edn, Better-Koehler Publishers, San Francisco

PILLAR 1

Thinking

01

Powerful brain, powerful leader

FRIDAY'S BUSINESS PLAN MEETING

It is Friday afternoon and everyone has shown up for the meeting. The whole week has been extremely demanding since the team was finalizing the business plan for the next year. Working until late at night was the norm. The last important details on key strategic issues are to be clarified in this meeting, and the tension in the room is evident. Everybody is physically and mentally exhausted and can't wait to get it over with. Although business planning has always been demanding, this year the pressure was higher than ever due to new market challenges. This business plan could make a big difference for the company by creating real growth, or it could perform below expectations, seriously damaging both the company's and the team's reputation. The team leader enters the room. She has always been a believer in leading by example, so she was the first to arrive at the office early in the morning and the last to leave late at night during this week. Her deep-rooted belief is that if they all worked harder and longer they would succeed. She takes a look at the agenda then at the strained faces around the table. The meeting begins.

Implementation arguments quickly evolved into confrontational lock-ins. Disagreements on strategy turned into nasty personal comments. And the widespread dissatisfaction progressed by the end of the meeting into several emotional breakdowns. Although the team ultimately managed to deliver a business plan that achieved its main goals, the team spirit that was so carefully and expensively built over the last year was irreversibly damaged. Relationships never recovered and two key people left the company within the next few months. Surprising as it may sound, the overall outcome of that Friday meeting had already been decided before it even started. Not by the people in that meeting. Not even by their leader. But by their brains.

How many times do leaders and managers face situations like the one described above? Formal meetings, important presentations, feedback sessions or simple decision-making regularly involve people who are strained and overworked. Regardless of the cultural stereotype of the all-powerful leader who is always on, constantly on top form and eternally in total control, science shows that brain strength has limits. If those limits are ignored, self-control can deteriorate fast, with dire consequences for all involved. Willpower is a key leadership characteristic and one that has to be carefully nurtured to produce desirable support for both clear thinking and better management of emotions. And it all starts in the brain.

The energy-devouring brain

The brain consumes more energy than any other organ in our bodies. According to a *Scientific America* article in 2008, the brain in total takes more than 20 per cent of the whole energy available to the body. Although it represents only 2 per cent of our body mass, it consumes one-fifth of the oxygen and one-quarter of the glucose (Foer, 2012). If it were an organization, this could mean that the brain represents a department with just 2 per cent of the total workforce, but one that demands 20 per cent of the company's financial resources. Naturally, every decision in this department would be taken with maximum efficiency and effectiveness. Prioritization would be the name of the game and every action would be painstakingly weighed by its impact on the survival of the whole organization. The brain behaves in exactly the same manner. It prioritizes body functions that ensure its survival, redirecting energy to those strategic areas from more exotic and luxurious functions such as analysis and forecasting, if and when needed. Going back to the organizational analogy, it is not very different from the fact that many companies, when coping with diminishing sales, focus on and commit more resources to those areas believed to be essential for the survival of the business. At the same time they take resources away from more elaborate and risky future projects and from expensive advertising campaigns.

Going a step further, neuroscience reveals that the brain consumes most of its energy on automatic maintenance systems rather than on executive, higher cognition functions. The brain consumes almost 90 per cent of its energy in a calm state, when people are not asked to do much thinking. So, it actually has a small portion of its energy available to devote to complex and cognitively demanding tasks such as managing a year-end meeting, debating the last details of a business plan and navigating diplomatically through the team's conflicting opinions. We go through our lives being confident that most of the brain's processing happens in the area we are aware of, our consciousness. This could not be further from the truth. Most of the power is used elsewhere, deep into our brains. According to estimations, the bandwidth of conscious awareness is 100 bits per second (Nørretranders, 1998), or even as low as 50 bits per second (Nease, 2016), out of about 10 billion bits of information entering the brain per second. This is no less than remarkable since it turns on its head a fundamental perception we causally hold for ourselves as species and as individuals. Actually, the part that we believe offers us our main competences and skills in developing ourselves and our careers, our analytical and conscious problem-solving part, receives just a fraction of the total energy and processing power. This means that we have to be extremely careful and strategic about managing this energy. Otherwise we will not have much to use in our demanding, challenging and exhausting daily information processing tasks.

A new theory has emerged on why the brain needs to use so much energy, constantly and mainly subconsciously, since this matter has been puzzling neuroscience for decades. James Kozloski (2016), a computer neuroscientist at IBM's Thomas J Watson Research Centre, developed what he calls a closed-loop brain model of neocortical information-based exchange, to explain all this extensive energy consumption by the brain. According to Kozolski, the brain constantly loops signals, which means that it is retracing established neural pathways repeatedly (Reynolds, 2016). In essence, over and over again, the brain runs signals through existing pathways to constantly make sure that three key functionalities are always working: sensory, what is happening, limbic, what it means to us, and behavioural, what is the appropriate response (Nield, 2016). This endless, and energy-devouring, process is evolutionarily beneficial since it gave us an advantage to survive and thrive over constant challenges from the environment.

The main question leaders need to answer, then, is how much power they will allow their brain to channel into its problem-solving section when they need to perform their best in such conditions. Are they going to create a situation where the brain will be starving for power and automatically reallocating it to more crucial and deeper structures for its survival? Or are they going to manage their working and living environments in such a way that will allow them to perform within maximum brain-power capacity when facing adversity? The first option represents the fast lane to bad decisions, angry responses and failed meetings. The second option represents the safe way for leaders to enhance their decision-making abilities and control over emotional reactions in an effective and meaningful way.

Before we start, though, we need to deal with a long-held myth about our brain's power and processing ability. One of the first brain facts that people mention in our lectures and keynotes, when asked what they know about the brain, is that we only use 10 per cent of it. This is not true. It is as absurd as saying that we use only 10 per cent of our arms. A healthy arm, although not used all the time, when needed uses all the necessary muscles and functionalities. The same goes for the brain. As Professor at Cambridge University, Simon Laughlin, says: 'the bulk of your neurons are relatively silent for long stretches of time, waiting to spring into action when activated. But they're doing so to remain energy efficient' (Richardson, 2019). So, brain cells rest until they are asked to spring into action and perform their necessary functionality. In a healthy brain, all cells are more or less ready to perform their tasks. Of course, popular movies, like Luc Besson's 2014 *Lucy* with Scarlett Johansson, which portrays Johansson's character, Lucy, gradually using more than 10 per cent of her brain and eventually reaching 100 per cent to become a god-like creature, do not help in abandoning the myth of the 10 per cent. Nevertheless, the movie was quite enjoyable.

The willpower muscle for leaders

Willpower and self-control are not infinite. Walter Mischel in his 2014 book *The Marshmallow Test* analytically describes and explains the famous Stanford marshmallow experiment, conducted by him in the late 1960s and early 1970s. Walter demonstrated graphically that when we are exposed to situations where we need to portray self-control we consume brain energy fast and thus perform much worse in the next task that needs willpower.

Children who demonstrated higher willpower when asked to restrain themselves from eating a marshmallow put in front of them performed worse in the next test on self-control that followed immediately after the marshmallow one. In contrast, those that showed less self-control and ate the marshmallow quickly in the first test performed much better in the second one. How could this happen? Isn't willpower something we either have or do not have? Results of the marshmallow experiment look counter-intuitive since we tend to believe that willpower is a key personality trait. Societies seem to separate people into strong and weak, whatever the situation, and people tend to stereotypically identify leaders as strong or weak based on specific prototypes that they have in their minds.

When we present the marshmallow experiment to our audience and ask them, students or executives, about the outcome of the *second* test, almost unanimously they respond in a predictable way: the children that showed higher willpower in the first test also did better in the second test. When we reveal the actual results, it usually takes them few minutes to accommodate this new view on willpower. This is another proof that it is not easy to overcome long-held beliefs about human nature and behaviour, even when scientific evidence is so strong. When they do come around to the idea that willpower is like a muscle that you can strain, they are able to explain past experiences better and see the future in a more confident and clearer way. They tend to recognize specific cases in their professional life during which they passed through an exhausting process for the brain that led them to think and act in a less effective way at a later stage.

ACTION BOX

Sit down, relax and then mentally go back to the last time that you have experienced a critical and demanding situation at your work that needed effort and fast decisions. Think about the things that consumed your brain energy and then think what happened afterwards. Consolidate your conclusions based on the arguments above and think what you need to be aware of next time. Do it. You will be amazed by the results.

The brain will utilize the power it has to help the person perform the desired task appropriately. This task could be a negotiation, a pitch presentation or a one-to-one coaching session. But being an energy-saving organ, when it is asked to consume a lot of energy and the overall energy levels fall it will re-prioritize and make sure that energy flows primarily where is needed the most. And this is always its survival and body maintenance centres. Thus, leaders need to be always aware of their brain's ability to distribute enough power to its executive part when is crucial for them to perform at their best. Otherwise, primal emotions, confusing thoughts and an inability to focus on the right things will take over and damage processes, relations and outcomes. *Ego depletion* settles in when your ego, as in your ability to direct your own actions and to perform effective self-control, is so tired and depleted of power that it cannot function properly anymore. In essence, it is not in control any more. You are not in control anymore. Primal brain functions have completely taken over your body. No leader wants to be in situations like that. Ever!

Ego depletion is found to involve at least four different parts of the human brain (Inzlicht *et al*, 2016):

- the part involved in planning and rule maintenance, called the dorsolateral prefrontal cortex (DLPFC);
- the part involved in inhibitory control and set shifting, called the ventrolateral prefrontal cortex (VLPFC);
- the part involved in conflict/error monitoring, called the dorsal anterior cingulate cortex (dACC);
- the parts involved in working memory maintenance and updating, within lateral parietal regions.

After reviewing existing neuroscientific research on the topic, Inzlicht and associates (2016) concluded that, 'from a social psychological perspective, each of these processes might be considered necessary, but not on its own sufficient, for self-control'. This makes self-control and its opposite, ego depletion, a complex neurological phenomenon with wide implications.

The good news is that the willpower muscle in leaders' brains can be improved in the long-term. The leading researcher of ego depletion, Roy Baumeister, in *Willpower: Rediscovering the greatest human strength* (Baumeister and Tierney, 2011), argued the benefits of strengthening our willpower muscle and the perils of not doing so. From all available recommendations from the science of self-control, our experience has shown that higher-level thinking, strong values and immediate feedback are three key strategies in companies for strengthening the leadership willpower muscle. Let us explain.

Higher-level thinking

Higher-level thinking is described as engaging in more abstract and creative thoughts or simply considering the big picture of a situation. This way of thinking, which is more conceptual, is related to 'why' questions and is the exact opposite of low-level thinking that asks more 'what' and 'how' questions. Experiments have found that higher thinking helps the brain perform better in willpower tests than low-level thinking. Although this scientific finding has profound practical effects for management and leadership, it is largely ignored in companies. Modern organizations seem to be obsessed with implementation and quick wins. 'Doing' is overly preferred over 'thinking' since fast market shifts require fast company responses. This leaves little space for engaging in true conceptual thinking and abstract contemplations. Even when such thinking is exercised, it is usually the privilege of people further up the hierarchy, if not reserved exclusively for the very top of the ladder. At the same time, though, companies demand that their front-line staff show total resilience and fight the daily battles in the volatile marketplace with advanced self-control and exemplary willpower. However, this can only happen when staff at all levels are encouraged to discuss openly the big picture, are exposed to 'why' questions and are motivated to reflect on the causes of things. The goal is to help people synthesize the whole rather than focusing on parts of the whole. In order to do so we need to recognize the contribution of high-level thinking to retaining our brain power. By facilitating this process we can enhance decisions and actions that are based on their willpower and they will exercise more self-control achieving objectives efficiently and effectively.

Strong values

Strong values have also been associated with increased willpower. Leaders wanting to be always focused and mentally strong need to develop meaningful and deeply held values. Hiring for attitude and training for skill is such a popular strategy, because people with the right work ethic will behave desirably most of the time, regardless of what they are faced with or how tired they are. They will simply have more energy to devote to their preferred behaviour than those that do not have values, or have questionable ones. Additionally, operating under strong personal values provides a blueprint to the brain of behavioural norms. So, in situations of low energy levels, when someone has worked very hard over the last few days and feels very tired, the brain will retreat to a habitual mode and replicate default behaviours to save power. These default behaviours, if guided by deep-rooted values, will align with aspirations and will reduce the possibility of undesirable reaction to external stimuli. Building strong organizational values that people really believe in can work in the same way and improve collectively the willpower in the company. But creating strong values goes beyond office branding and in-house mantras. It should run in the DNA of a company from the top to the bottom. Values have to become the absolute yardstick for measuring all behaviours and all performances. Only in that way will the brain learn that these are absolutely necessary for survival and relegate them to its region of automatic behaviours. This will give it the necessary energy for them in cases where energy is scarce. In our experience, values become an integrated part of corporations' collective mind when following the ABC framework for infusing organizational values.

THE ABC OF CREATING ORGANIZATIONAL VALUES

A is for authority. Organizational values are implemented and transited in a top-down manner, from positions of high authority to positions of low authority and not the other way around. This is because the brain of people lower in the hierarchy will subconsciously copy and paste the actual behaviours of people seen as having more power than them.

B is for barometer. Organizations need to measure periodically the status and maintenance of values among their members. The two main variables to be measured are the *importance* of each value and the *performance* in each value. That is, we need to measure what is the potential impact of each value in our business (importance of the value) and how much each value drives behaviour in the company (performance of the value).

C is for consistency. Each value should not be used just for internal PR purposes and for creating a nice brand story. Every value should be highly integrated in all organizational processes and procedures in a formal way that is clear to all. This can enhance their maintenance by quickly creating appropriate habits.

Immediate feedback

Feedback provides tangible checkpoints for the brain to adjust behaviour (more information and suggestions about feedback can be found in Chapter 8). Absence of self-control usually makes a person more agreeable, more sensitive to impulses or even more agitated. Immediate feedback can help in restraining a bad behaviour and even in re-evaluating the situation, leading to corrective actions. For example, if you receive formal feedback that you need to work on your self-control in order to be considered for a promotion, it can have a strong influence on your behaviour the next time you are in a meeting. Similarly, you can ask a trusted member of the team to informally signal to you when the meeting is getting hotter, if you feel you are not always calm enough yourself to detect it. On a broader scale, developing an organizational culture that is transparent and supportive, and with widespread processes for delivering fast and accurate feedback, can help employees' brains receive the right signals at the right time to adjust their behaviour and thus to show more self-control and willpower. The more important the signal, the more the brain will give it the energy and attention it needs to deal with it. So, are we sure we are receiving the right feedback at the right time? Moreover, are we giving this feedback the right attention and weight to assist our brain to behave accordingly? If not, then the power we need to respond strongly and affirmatively to the signal might simply not be there.

The science of willpower provides significant insights on what it takes for a leader to have a strong brain, but it does not tell the whole story. In order to have a more complete picture on brain power and make sure we supercharge our brain to deal with complexity, we need to avoid the burnout syndrome.

The burnout syndrome

So you finally took that long-awaited holiday to an exotic destination! It was well-deserved, since you have been working non-stop for the past year with multidisciplinary and multinational teams on a crucial restructuring project. The project was very exhausting for its entire duration. You often daydreamed of this holiday and the time you will spend away from everything, just enjoying the sun and the sea. You would forget everything for a while and return to the office as good as new, super-ready for the next task. The problem is that as soon as you got back to the office after your holiday, and regardless of the physical rest, the mental rejuvenation and the emotional recharge you experienced, one week into the job you started feeling as tired as ever. Let's be honest. A lot of us had this strange feeling and, even worse, found it very difficult to motivate ourselves to explain it, thinking that we need to rediscover our old performative self. However, this is not how it should work. You got your time off, thinking of big ideas or of no ideas at all. What's going on? Where do the tiredness and all the negative feelings towards the job come from? Chances are, they come from the burnout syndrome that decreases your ability to perform as well as you always wanted.

Feeling chronically tired, both physically and mentally, is not uncommon in modern organizations where new problems arise every day and the lessons learned yesterday do not always apply to the challenges of tomorrow. But the burnout syndrome does not depend only on tiredness and exhaustion. Again counter-intuitively, it also depends on core job conditions that, if not taken care of, can cause your brain to disengage and perform below everyone's expectations and mainly yours. Can you be a true leader when suffering from burnout? We don't think so either. How can a leader inspire and influence others when being constantly exhausted, cynical or feeling ineffective, or all of the above together? Because these are the main symptoms of experiencing burnout. And how can a leader lead a team that is burned out too, experiencing these symptoms for months and months in a row? Almost impossible.

At times in our consulting and coaching careers we have encountered companies with widespread burnout feelings. Departmental directors, middle managers and front-line staff alike complaining constantly of being devoid of energy, having to do too much, too fast, with more tasks being added daily and without any mercy. Even worse, they are often not in a position to describe effectively the situation that they face. They do feel and understand what they are going through daily but words do not seem to be enough to communicate this. So they ask for help at any given opportunity. In such cases the corporate culture seems to be like an energy vampire that depletes its employees of their mental powers on purpose. Digging below the surface, a leader can very easily uncover the true causes of burnout and take immediate actions to make sure that it no longer influences his performance. By doing so, leaders can confront and reduce any trace of cynicism, boost their self-confidence and feel more empowered to engage themselves and others into achieving their goals.

Burnout is a serious organizational issue. A Gallup survey revealed 23 per cent of employees felt burnout actively while another 44 per cent felt it occasionally, leading to significant employee disengagement and turnover effects (Borysenko, 2019). In fact, it is so prevalent in workspaces around the world that in 2019 the World Health Organization (WHO) included burnout in its 11th Revision of the International Classification of Diseases as an occupational phenomenon. Although not a disease in itself but a symptom, WHO explains that burnout is 'conceptualized as resulting from chronic workplace stress that has not been successfully managed' (WHO, 2019). The fact that burnout is mostly about management's ineffectiveness in resolving chronic pressure on employees illustrates the central role of leadership in creating a nurturing environment that will detect and address burnout in a timely and decisive way. And the first step towards creating such an environment is for leaders to make sure they are not burned out themselves.

A prominent case-in-point of a burned-out leader is Elon Musk, the CEO of Tesla and SpaceX. On 7 August 2018 Musk tweeted that he had secured funding to take Tesla private at a specific share price (Rapier, 2019). First, this was not actually the case, and second, he should not be tweeting such information without the appropriate consultations and discussions with his team, partners and investors. As a result, Musk stepped down as Tesla's chairman, and both he personally and the company paid hefty fines. In an interview a few days later Musk admitted that the previous year had been the most painful and difficult in his career, and that he had been working 120 hours a week prior to the tweet (Gelles *et al*, 2018). Musk admitted that 'there were times when I didn't leave the factory for three or four days – days when I didn't go outside. This has really come at the expense of seeing my children. And seeing friends' (Gelles *et al*, 2018). Musk's situation showcases that depletion and burnout are not reserved for employees at the lower and middle levels of the hierarchy. They concern everyone, from an intern to the CEO. If burnout is not diagnosed early and treated successfully, it will lead to mistakes, a fall in productivity, erratic actions, withdrawal, disconnection, conflicts, absenteeism and other unwanted organizational behaviour.

Physical exhaustions should be addressed in various ways. Holidays can be a good solution, but daily and weekly doses of relaxation can help much more. A leader has to be aware of the state of being 'high' that adrenaline can bring you to when working long hours for an extended period of time. Although it feels momentarily good, it does so to hide pains and inefficiencies of the body and mind. It makes you keep going when you probably need to take a break and re-focus. Creating a working environment that allows for breaks and relaxing moments is vital for dealing effectively with burnout. The impact of recovery or decompression sessions cannot be stressed enough. However, the deeper causes of burnout are harder to deal with. But they should be dealt with decisively if we want to calibrate our brains for leadership.

The keyword in understanding and preventing burnout is *mismatch*. When mismatch is experienced for a period of time without being resolved, it can lead to one or more symptoms of burnout. This mismatch can occur in many forms:

- increased complexity of a problem and limited time to resolve it;
- difference between job description and actual skills;
- nature of task/organizational culture against personal values;
- high expectations to perform and lack of available resources;
- responsibility for a task but absolute lack of control over it;
- pressure to perform accompanied by lack of appreciation by the company;
- very close structural relationship with very bad quality of relationship (supervisor over supervisee, for example);
- variance between organizational 'talk' and 'walk' – managers saying one thing and doing another;
- newcomers faced with highly unpredictable jobs;
- perceived seriousness of work and impoliteness of co-workers.

The list unfortunately can go on forever, and managers and employees around the world are welcome to add their own mismatches here. Even worse, many are faced with a combination of the above mismatches, which causes multiple symptoms of burnout to manifest simultaneously. All in all, mismatch creates a very toxic environment for sustaining leadership qualities. So, what do we do?

ACTION BOX

Think of a couple of complex issues that you needed to deal with. Then try to find the specific mismatch that made you feel exhausted and unwilling to continue performing. What are your conclusions? What can you do next time to avoid similar mismatches? Remember, the important point is for you to identify what to avoid doing next time.

Our brain is a social organ. The fact that the worst punishment in prison is isolation proves the point that our brain can deal with many things, but not with absence of interaction. Our brain grows and becomes better through social interaction. Without meaningful and constant interaction in the early stages of our lives we can end up with serious mental issues and deficiencies. Our brain develops and learns mainly through external stimuli and our social surrounding plays a very important role. Although the relational part of fine-tuning the leadership brain will be discussed in later chapters, forging the right social interactions at work is considered to be the universal medicine for burnout.

Through our extensive experience in internal communications advisory we have identified a number of occasions where improved information exchange internally reduces the mismatches mentioned above. For example, strategically designed, two-way internal communications can help leaders listen to their teams and transmit the right information to the right people. Explaining things comprehensively, and discussing company values, objectives, processes and results are taken lightly by so many companies, even though this is the easiest way to deal with mismatches. Information exchange through open communication is necessary for any leader to fight burnout both individually and collectively. Even at the micro level, every good relationship within a company is based on timely, accurate and interactive communication between two people or within a team. This is really unavoidable. Furthermore, a smile can go a long way. Treating people with respect and spreading kindness around create the right conditions for reducing the impact of burnout on others, and in return on you. Last but not least, prioritizing the right tasks within your day and in your team, and making sure everyone understands what is happening and why, are good ways of decreasing mismatches. Decreased self-control and the burnout syndrome can damage your leadership ability for good. Knowing how to spot symptoms and how to treat them in a timely manner will help you drastically improve your leadership brain. Nevertheless, there is one more condition that threatens your brain strength and that needs to be addressed by the modern leader. Multitasking is probably the trickiest of the three because many managers and directors we meet in companies are convinced that it is actually a positive characteristic of leadership. Multitasking hides more dangers than we think.

Multitasking yourself into cognitive load

Complexity, in the form of multiple, unpredictable and dynamic forces shaping our everyday business reality, has increased drastically over the last decades. Technology and global competition are just two of the elements of the external environment that change in a fast and unforeseen way. If shifting demographics, global politics, emerging scientific insights, demanding customers and the fluid state of world economics are added it is not difficult to understand the unprecedented pressure that modern leaders experience. Results, new product pipelines, talent management, value to shareholders and quick wins are areas where they are expected to perform spectacularly, often without having the luxury of hiring more personnel. Consequently, many people in companies believe that multitasking allows them to deal successfully with this complexity by simply performing more tasks at the same time. Does it work?

It doesn't. Regardless of the popular belief that multitasking is an impressive skill of super-efficient people, it actually decreases the brain's power, weakening leaders. This is because multitasking is related to higher cognitive depletion. The more tasks you put on yourself to do simultaneously, the more pressure you add to your working memory. Ultimately, the higher the *cognitive load* in your brain, the faster it gets depleted from its available power. This should not be a surprise to managers since the negative relation between performance and overload seems to exist in many other types of systems. An overflow of simultaneous input causes congestion: from electricity grids to roads and from concert halls to rivers. Why did we think that our brains are different?

It is true that adults have a higher capacity for managing different tasks at the same time than children and elderly people. It is also true that education and experience can generally enhance the capacity for multitasking. But up to a point. The information load we generate as a society today has increased at an unbelievable rate, and much faster than the brain's ability to deal with it. As reported in the technology portal techcrunch.com (archived at https://perma.cc/C7UX-8FXL), Eric Schmidt, the Google CEO, said at a technology conference in 2010 that the amount of information we collectively produce every two days is equal to all the information produced by humanity up to 2003, mainly boosted by the extreme speed with which user-generated data are posted online. This startling claim should alarm everyone in organizations since our current approach to multitasking needs to be urgently revised. Instead of treating multitasking as the antidote to increased complexity, pressure and information, we should consider it as one of its diseases. Multitasking is not helping our leadership brain take better decisions. It does not allow it to focus on the right issues and inspire people. It rather creates an illusion of efficiency fuelled by an inner urge to feel proud of how much is accomplished in a limited amount of time. Multitasking is really not about efficiency. It is about sensation and about masking shortcomings in core leadership characteristics. We also need to stop thinking of leadership practice as another task. It is *not*. A study by Sanbonmatsu *et al* (2013) provides a number of amazing insights that should challenge the positive attitude towards multitasking in companies. The study found that people who considered themselves to be great multi-taskers actually had an inflated opinion of themselves. They performed very badly in multitasking tests. Apparently, they even did worse than people who did not consider themselves multitaskers and did not engage much in multitasking in their everyday lives. That is an amazing finding. The more we multitask and the more we consider ourselves as fantastic multitaskers, the less effective we are in reality. If this is the case, then why do people do it? Often, multitaskers are thrill-seekers hunting for the excitement that achieving many tasks could bring. People labelling themselves as serial multitaskers are more impulsive, embracing whenever possible the rush of doing many things at the same time. *Doing* does not mean *doing right*, though. It might feel good and exhilarating but it does not make us better leaders. Leadership is not a task for a multitasker. Leadership is about communicating, showing the way, motivating, supporting and developing, and these are not traditional job tasks. They are, or they should be, daily behaviours of managers. By doing various things in parallel and trying to achieve multiple targets simultaneously, we may end up with serious losses in the leadership game.

Sanbonmatsu *et al* (2013) went on to reveal that self-proclaimed multi-taskers demonstrated important flaws in controlling the executive part of their brain. The executive part controls our impulses, directs attention and focus, deals with our working memory and manages our problem-solving and task completion functions. Cognitive load can inhibit all these, making leaders weaker and incapable of fulfilling their influential role. How can we trust and follow a leader who demonstrates low attention, confused working memory and limited problem-solving? We cannot. This is why multitasking should be approached with extreme caution. Multitaskers are less able to manage distractions and direct enough focus on a single task at a time. However, with increased complexity and the unmanageable rate of information flow, distractions and noise are here to stay. Businesses need leaders that will not jump impulsively every time a new direction appears. Quite the contrary, they need people who can navigate with conviction in the sea of challenges and lead others in the right direction. With the limited amount of energy in the brain available for the executive part, it is extremely risky to weaken its important executive functions because of too many tasks processed at the same time. True leaders are those that use their brain power wisely and selectively to make sure that they have enough for focus, problem solving, impulse inhibition and reasoning. Otherwise, those functions will receive less power than needed to work properly, resulting in confused, disoriented and frustrating behaviours. Self-control suffers greatly.

LEARNING FROM REAL LIFE: THE MULTITASKING MANAGER

We specifically remember a case where we were advising a business development director of a large retail company in the south east European region. With global experience and many international successes under his belt, he was confident about himself and his team. However, when shockwaves from the 2008 economic crisis hit the region he had to deal with many 'fires' at the same time. This was a new situation for him. Therefore, he tried to take on more responsibilities, attempting to confront increased complexity with increased multitasking. He was working more hours with multiple tasks on his daily agenda. He pressurized himself and others, and started creating more problems than the ones solved. It took him some time, though, to understand that multitasking was part of the disease and not the solution. This was because spreading his attention to multiple locations and hot issues simultaneously was creating excitement and an overblown feeling of short-term accomplishment. It was obvious to us that the quality of his attention, prioritization and decision-making was deteriorating fast. Fortunately, an early mishap and the analysis of the potential bigger consequences of his multitasking drive made him reconsider his approach. With our help, he:

- delegated more by empowering his closest people to take some of the crucial decisions as well;
- focused on one core issue per meeting by increasing the number of meetings but decreasing drastically their duration;
- organized his work in a way that allow efficient grouping of related tasks;
- set time aside to reflect on and evaluate new information and new tasks before moving to execution;
- increased transparency of workflow across the team so everyone was aware of what was happening at any time;
- used feedback better from both inside and outside the company after every task was completed to make sure that things moved as planned;
- took corrective actions first before moving to a whole new task;
- avoided unnecessary distractions by minimizing exposure to non-crucial sources;
- created quiet sessions within the week for concentrating better on the task at hand;
- adopted a new weekly schedule that allowed for surprises and reprioritizations as much as possible.

The company weathered the crisis better than the competition and managed to sign a few more lucrative deals in the process. The company did well because the above ideas shifted the whole cultural environment inside the company, making other people follow similar routes since he was doing so. And the leadership game was won.

An interesting detail from this case is the fact that at the start of the high-pressure period our client was actually doing very well with multitasking. The study by Sanbonmatsu and associates did show that non-multitaskers performed better in multitasking when asked to do it than those proud of their multitasking abilities. This means that we will not necessarily fail in multitasking if there is no alternative, for a short period of time. However, if this becomes a habit and we get hooked on the sensational 'high' of multi-tasking then we endanger both ourselves and our companies. Having said that, there is a tiny part of the population that could be characterized as *supertasking* due to their ability to multitask without the associated cognitive shortcomings of multitasking. Neuroscience still cannot explain the phenomenon, so to consider ourselves as supertaskers without close scientific observation is not only unreliable, but dangerous. But let us see some more facts.

A study in multitasking by Bachmann *et al* (2019) shed some more light on how multitasking can be done in a more successful way. The researchers studied students during a week and found that the students multitasked for about 41 per cent of the time. However, not all multitasking was the same. The impact on participants' perception and feelings towards the various tasks depended heavily on the degree of autonomy in performing a task. When additional tasks were imposed, the impact of multitasking on participants' well-being was negative, while when additional tasks were more autonomous, the impact on well-being was positive. This means that when we are asked to perform more than one task at a time, if we feel coerced and controlled in doing so, the effect of multitasking will be worse. However, if the additional task is one that we welcome and even initiate ourselves, with a high degree of autonomy and personal involvement, then the effect of multitasking can even be positive.

Having in mind that multitasking long term should be avoided, this study helps in understanding how to perform the necessary amount of multitasking in a better way. Increasing the sense of control by empowering people to choose their tasks and manage their time can lead to better multitasking when this cannot be avoided. High engagement and personal emotional investment in what we do have the ability to mitigate the negative effects of multitasking if it is necessary. On the other hand, loading people's to-do lists in a top-down, authoritarian way, and giving them little to no control over performing additional tasks 'because they have to' makes multitasking a modern chimaera of companies worldwide.

Resilience is the answer

Complexity, and the sheer amount of challenges facing the modern leader, will not slow down anytime soon. Leaders will need to navigate and operate in highly demanding environments having the same brain, as a species, as thousands of years ago. This means that our ability to deal with large amounts of information, dynamic decision-making and multitasking, which are key drivers of ego depletion and burnout, will need to improve too. In addition to the ways presented in this chapter so far for achieving this, there is one more concept we would like to discuss: resilience.

Our brains constantly develop and revise strategies for dealing effectively with the environment. Based on where we were born and the experiences we went through during our lives, our brain has learned to fully immerse itself in or to deflect emotionally compromising situations. The degree to which the brain sustains the emotional hit from a negative situation and the speed by which it recovers from it is called resilience. Practically speaking, if your thinking process remains relatively unaffected and your willpower strong after negative news such as a significant drop in sales, an important vendor's delayed delivery and the defection of a strategic partner to a competitor then your resilience is high.

Resilience as a term is frequently used in relation to engineering. It is interesting to examine it from that perspective in order to reveal its inner meaning. Professor of Integrated Systems Engineering and Cognitive Psychologist David D Woods defines resilience as 'understanding how the system adapts and to what kinds of disturbances in the environment' (Woods, 2006). According to Woods, its essential characteristics are:

- *Capacity:* The size or kinds of disruptions the system can absorb or adapt to.
- *Flexibility:* The system's ability to restructure itself in response to external changes or pressures. The opposite is stiffness.
- *Margin:* How closely to its limits the system can continue operating.
- *Tolerance:* How a system behaves when reaching a boundary (fast vs slow collapse).

In the organizational context, those essential characteristics can be translated as follows: how many 'hits' (negative news, bad decisions, personal conflict with colleagues, etc) you can take and what kind of 'hits' have a bigger effect on you; how flexible are you when you get a negative 'hit' and how do you compose yourself; how close to your limits can you operate normally; if you reach your limit, how fast or slow do you regress? These are difficult questions, and their answers are not always known to us, but they are important ones that need to be addressed.

In work, psychology resilience has a very similar definition to that used in engineering. In an overview of the field, Fletcher and Sarkar (2013) defined resilience as 'the role of mental processes and behaviour in promoting personal assets and protecting an individual from the potential negative effect of stressors'. Resilience is like the brain's immune system for challenging situations. Having said that, a resilient person is not the one that is not affected at all by the situation. This would probably be a sign of indifference, disassociation and disengagement. The resilient person receives the negative 'hit', experiences it, and then reacts and 'bounces back'. Various components of personal resilience have been suggested in relevant literature such as: degrees of coping, personal flexibility, sense of agency and purpose, positive engagement in daily life, emotional regulation and biometrics of physiological buoyancy, such as heart rate variability (Zautra *et al*, 2010). Since resilience is a process, it is not static, but it develops over time.

Studying brain network architecture, Ohashi and associates (2019) found that resilient individuals had less interconnected brain sections than non-resilient ones. This was particularly pronounced in the right amygdala, which is the main fear centre in the brain. This means that resilient brains have more isolated brain sections in order not to allow widespread distribution of a strong negative reaction throughout the brain. This is a remarkable finding. It adds to previous studies in both animals and humans showing that resilient brains have learned to use neuro-plasticity, the brain's ability to physically change, to address challenges and promote right behaviours to their advantage (Russo *et al*, 2012). In contrast, non-resilient brains have learned, often due to earlier trauma, to be more sensitive, or overly sensitive, and to prioritize threats and negativity. Indeed, resilience is the process in which the brain engages actively in coping mechanisms to return to the normal situation, or its homeostasis, and not overemphasizing the risk by making it dominant in extensive neural networks.

Resilience is also a well-applied concept in leadership studies. In particular, leadership resilience, along with other positive psychological capacities like, hope, optimism and confidence, are considered to be psychological capital (PsyCap) for leaders (Luthans *et al*, 2006). Peterson *et al* (2008), by collecting and analysing data regarding PsyCap of 55 business leaders from a variety of industries using survey techniques, managed to categorize leaders into low and high PsyCap. Then by using a well-established neuroscientific research method called electroencephalography (EEG) they compared the two groups and found that high PsyCap ones tended to show greater activity in the left, prefrontal cortex, confirming that resilience is related to this part of their brain. According to the same study, resilient leaders have a more reliable view of reality, being more logical in analysing situations and taking action. In addition, Maulding *et al* (2012) found that resilience is related to effective leadership. Their study found a strong correlation between resilience, emotional intelligence and successful leadership performance. Similarly Steward (2014) reveals that resilient leaders show a series of capabilities like emotional intelligence and energy, but also a series of self-oriented aspects like well-being, strong purpose and self-confidence. These studies and many more suggest that resilience is not a state of being, but a complex process of adaptation and growth within a particular context (Southwick, *et al*, 2017). A resilient leader does not just manage to survive in a challenging business environment – they embrace challenge and thrive in this environment. So, the question now is how can leaders develop their resilience? What are the aspects that they can enhance?

According to the American Psychological Association (APA, 2019), these are the 10 main ways to increase resilience:

1 *Make connections*. Create and engage in relationships that are meaningful, by receiving and giving assistance to people close to you.

2 *Rethink crisis*. Instead of looking at crises as catastrophic, one-off events, try to view them as situations that have happened before to other people who are still around.

3 *Embrace change*. Accept the fact that everything changes and that this might be a good thing in helping you focus better on what matters most.

4 *Set goals*. Create manageable and realistic goals and move towards them in a feasible step-by-step approach.

5 *Take decisions*. Engage with challenges by taking decisions and moving into action.

6 *Look inside*. Identify situations where you felt slightly better and try to adopt a similar mindset when problems manifest.

7 *Increase confidence*. Learning to trust your intuition and instincts can create a more positive view of yourself.

8 *Think long-term*. Dissecting a negative situation by looking at its past, present and future is key to demystifying it.

9 *Be positive*. Focus on the positive aspects of your life, or of a challenging situation, and visualize a better outcome.

10 *Take care*. Try to enjoy life and work more by doing the things that have a higher positive impact on you.

When your brain maintains optimum power levels, the ability to increase your resilience is at its best. A tired brain is rarely, if ever, a resilient brain.

Keep in mind

Multitasking, ego depletion and the long-term syndrome of burnout are serious threats to the modern leader and the practice of leadership. Brain power is crucial for leadership since it is directly related to increased self-control, meaningful engagement, effective task completion and resilience. Thus, a powerful brain creates a powerful leader and not the other way around. You need to embrace the fact that the guardian of your brain's power is you. So, choose to protect it and to direct its energy where is needed the most.

BOOST YOUR BRAIN: LEARN TO DISTINGUISH TASKS FROM LEADERSHIP

Make a list of tasks that you want/need to accomplish almost daily in your job. Then make another list of your actions as leader. Think what you are doing to accomplish the items on the first list and what you do in order to behave as leader. Then think which of these tasks are really critical and which leadership actions are equally critical. Think how much time you dedicate on the task side and then how much time you dedicate on leadership actions. Do you leave your brain enough power in both groups? How can you improve your brain power reserves based on what you learned in this chapter? Consider your conclusions and use them next time.

References

APA (2019) Building your resilience, www.apa.org/topics/resilience (archived at https://perma.cc/E324-T5GY)

Bachmann, O, Grunschel, C and Fries, S (2019) Multitasking and feeling good? Autonomy of additional activities predicts affect, *Journal of Happiness Studies*, 20(3), pp 899–918

Baumeister, FR and Tierney, J (2011) *Willpower: Rediscovering the greatest human strength*, Penguin Group, London

Borysenko, K (2019) Burnout is now an officially diagnosable condition: Here's what you need to know about it, www.forbes.com/sites/karlynborysenko/2019/05/29/burnout-is-now-an-officially-diagnosable-condition-heres-what-you-need-to-know-about-it/#45495e092b99 (archived at https://perma.cc/Q7RY-KLFM)

Fletcher, D and Sarkar, M (2013) Psychological resilience: A review and critique of definitions, concepts, and theory, *European Psychologist*, **18**, pp 12–23

Foer, J (2012) *Moonwalking with Einstein: The art and science of remembering everything*, Penguin Books, London

Gelles, D Steward, JB, Silver-Greenberg, J and Kelly, K (2018) 'Worst is yet to come': A fraying, exhausted Elon Musk confronts a fateful tweet and 'excruciating' year, https://business.financialpost.com/transportation/autos/a-fraying-exhausted-elon-musk-confronts-a-fateful-tweet-and-an-excruciating-year (archived at https://perma.cc/JMM6-CA5T)

Inzlicht, M, Berkman, E and Elkins-Brown, N (2016) The neuroscience of 'ego depletion': How the brain can help us understand why self-control seems limited, in *Social Neuroscience: Biological approaches to social psychology*, 1st edn, ed E Harmon-Jones and M Inzlicht, pp 101–23, Routledge, New York

Kozloski, J (2016) Closed-loop brain model of neocortical information-based exchange, *Frontiers in Neuroanatomy*, **10**, p 3

Luthans, F, Avey, JB, Avolio, BJ, Norman, SM and Combs, GM (2006) Psychological capital development: Toward a micro-intervention, *Journal of Organizational Behavior: The international journal of industrial, occupational and organizational psychology and behavior*, **27**(3), pp 387–93

Maulding, WS, Peters, GB, Roberts, J, Leonard, E and Sparkman, L (2012) Emotional intelligence and resilience as predictors of leadership in school administrators, *Journal of Leadership Studies*, **5**(4), pp 20–29

Mischel, W (2014) *The Marshmallow Test: Understanding self-control and how to master it*, Corgi, London

Nease, B (2016) *The Power of Fifty Bits: The new science of turning good intentions into positive results*, HarperCollins, New York

Nield, D (2016) Neuroscientist says he's solved the mystery of why our brains use so much energy, www.sciencealert.com/scientists-think-they-ve-figured-out-why-the-brain-uses-up-so-much-energy (archived at https://perma.cc/65JQ-RZZN)

Nørretranders, T (1998) *The User Illusion: Cutting consciousness down to size*, Penguin Books, New York

Ohashi, K, Anderson, CM, Bolger, EA, Khan, A, McGreenery, CE, and Teicher, MH (2019) Susceptibility or resilience to maltreatment can be explained by specific differences in brain network architecture, *Biological Psychiatry*, **85**(8), pp 690–702

Peterson, SJ, Balthazard, PA, Waldman, DA and Thatcher, RW (2008) Neuroscientific implications of psychological capital: Are the brains of optimistic, hopeful, confident, and resilient leaders different? *Organizational Dynamics*, **37**(4), pp 342–53

Rapier, G (2019) Elon Musk's infamous tweet declaring he had 'funding secured' at $420 per share to take Tesla private was sent one year ago today. Here's everything the company's faced in the year since, www.businessinsider.com/elon-musk-tesla-private-one-year-anniversary-2019-8?r=US&;IR=T (archived at https://perma.cc/SP46-5XDL)

Reynolds, E (2016) Why does the brain use so much energy? www.wired.co.uk/article/why-brain-uses-so-much-energy (archived at https://perma.cc/Z3V8-L6B2)

Richardson, MW (2019) How much energy does the brain use? www.brainfacts.org/brain-anatomy-and-function/anatomy/2019/how-much-energy-does-the-brain-use-020119 (archived at https://perma.cc/5UVM-56BL)

Russo, SJ, Murrough, JW, Han, MH, Charney, DS and Nestler, EJ (2012) Neurobiology of resilience, *Nature Neuroscience*, **15**(11), pp 1475–84

Sanbonmatsu, DM, Strayer, DL, Medeiros-Ward, N and Watson, JM (2013) Who multi-tasks and why? Multi-tasking ability, perceived multi-tasking ability, impulsivity, and sensation seeking, *PLOS ONE*, **8**(1)

Siegler, MG (2010) Eric Schmidt: Every 2 days we create as much information as we did up to 2003, http://techcrunch.com/2010/08/04/schmidt-data/ (archived at https://perma.cc/4P9M-LKXN)

Southwick, FS, Martini, BL, Charney, DS and Southwick, SM (2017) Leadership and resilience, in *Leadership Today*, ed J Marques and S Dhiman, pp 315–33, Springer, New York

Steward, J (2014) Sustaining emotional resilience for school leadership, *School Leadership and Management*, **34**(1), pp 52–68

Swaminathan, N (2008) Why does the brain need so much power? *Scientific American*, 29 April

Walter, M (2014) *Marshmallow Test*, Transworld Publishers, London

WHO (2019) Burn-out an 'occupational phenomenon': International Classification of Diseases, www.who.int/mental_health/evidence/burn-out/en/ (archived at https://perma.cc/76T6-HGPK)

Woods, DD (2006) Essential characteristics of resilience engineering, in *Resilience Engineering: Concepts and precepts*, ed E Hollnagel, DD Woods and N Leveson, pp 21–34, Ashgate Publishing, Burlington

Zautra, AJ, Hall, JS, and Murray, KE (2010) A new definition of health for people and communities, in *Handbook of Adult Resilience*, ed JW Reich, AJ Zautra and JS Hall, pp 3–29, Guilford Press, New York

02

Clear mind, strong direction

THE PATTERN HAS BEEN REVEALED TO ME

The moment of truth. The CEO is about to announce the yearly numbers in front of the whole team. It has been very difficult recently and the company is struggling to come up with competitive ideas. New players are eating into its market share like never before. The team is disappointed with the third quarter's results and is expecting better sales figures in the last quarter. This is because the CEO, and admired leader of the company for so many years, has taken a series of bold decisions to reverse the negative trend, decisions they all supported since his charm and conviction are as strong as ever. He has been very confident about improving results and has the right insights to prove it.

At the beginning of the year he engaged a research agency to help him understand the situation better. It was the first time in his long career that he could not read the market as effortlessly as he always did and he wanted to go deeper. Together with the agency and a few trusted directors, they designed an ambitious study that aimed to measure all critical aspects of the business, internally and externally. The study took months to design and implement but six months ago, when the numbers came in, it became very clear to him what needed to be done. Although his team struggled to put everything together and see the big picture when the complicated and multilayered results were presented, he immediately got it. He felt young again, rejuvenated by his ability to see through all the graphs and capture the meaning of it all. He saw the pattern clearly. Full of confidence, he dictated to his team what they had to do as of the following day. When the procurement director expressed her reservations, he and the other directors calmed her down and reassured her that it was all going to be OK. They could all see it by now, anyhow. The pattern was finally there and it felt so good. When the CEO entered the room to announce the results at the end of the year his facial expression said everything: negative numbers do not go away. It took two years for the company to recover and the people to understand.

How many times do managers and leaders experience the invigorating feeling of pattern recognition? How often colleagues shout 'a-ha' in a meeting, having just conceived the ultimate truth from the numbers on the screen? And how many times have those patterns been as far as possible from reality? The brain loves patterns. It does so because it makes it feel secure in situations of high uncertainty. If there is danger out there and you are not sure of what to do you are not going to move much. Not moving is more dangerous than moving so the brain will try to find clues to work out fast the best option and to go for it. And it will work it out regardless of the quality of the clues. So, problems arise when the selected course of action is based on the wrong understanding of the situation. Although happy to move, we can end up in a much worse situation than before the action. Leaders should take decisions while being aware of such brain traps, in order to make sure they are not fooled by them. This will prevent their actions from fooling others as well.

The evolution of the survival-obsessed brain

We go about our lives being conscious. Well, most of the time, that is, and when we are not asleep in dreamless state. Being conscious gives us a feeling of being in control since we can observe our thinking and see when and how we make decisions. It is not by accident that our species is called *Homo sapiens sapiens* or 'the wisest of the wise'. We are the only species with advanced self-awareness and language capabilities. Although other mammals like chimpanzees, dolphins and even magpies are also self-aware, humans are unique in the animal kingdom in engaging in sophisticated and elaborative thinking. This uniquely superior capacity, not just to think but also reflecting on our own thinking, has been long hailed as the pinnacle of what makes us human. 'I think, therefore I am' said Descartes in the 17th century, while much earlier Aristotle talked about rationality (*logos*, meaning in Greek language logic) being a distinct human ability and thus humans being rational animals. Even art glorifies our thinking processes, as best embodied by the famous bronze statue *The Thinker* by Auguste Rodin which first appeared at the turn of the 20th century. Somehow, it has become an axiom that thinking is controlling the human brain, or has the ability to do so, since this is what makes humans human. Thinking, represented by the executive part of our brain, became the absolute benchmark for human existence, the absolute skill and the absolute driver. Neuroscience, though, says otherwise.

The absolute purpose of our brain is not thinking. As we usually like to say in our lectures, the ability to use thinking exclusively and every time is in reality wishful thinking. Thinking is the means to an end, and the end is survival. And if survival is the aim then the brain will use all its functions to support it in an integrative and definite way. The brain assesses a specific situation and, based on the urgency and importance to survival, it will engage different functions to fit its purpose. Thinking is rarely a priority for the brain when reacting to a situation. This is because the executive brain functions, the ones that control emotions, forecast the future, estimate consequences and do calculations developed much later in evolution. Approximately 1.9 million years ago, when the *Homo* species first appeared, the brain started developing more sophisticated frontal lobe areas related to advanced language skills. So far, so good. However, brains did not first appear in humans. Brains as concentrated nerve cells controlling bodily systems and behaviour exist in nature much earlier. In the anatomy of the human brain, the older structures, usually referred to as the hindbrain, and the midbrain, usually referred to as the limbic system, exert much more influence on our behaviour than the newest addition, the neocortex or forebrain. Whenever we ask people in our seminars to draw the brain they actually draw the neocortex, which seems to be the image most of us associate brains with. The real influence of the brain though comes from deeper inside.

The hindbrain, or, as it is also called, the reptilian brain, represents the more basic functions of our brain, like controlling balance, heart rate, breathing and temperature and is found in the brain stem and cerebellum. It is very similar to a reptile's brain. As explained by Dan Hill (2010) in his book *Emotionomics: Leveraging emotions for business success*, this type of brain has existed for approximately 500 million years in nature. The limbic system, or border system, between the brainstem and cerebral hemispheres includes functions controlling learning, memory, motivation and emotion.

This type of brain has existed for approximately 200 million years in nature. Last and probably least, the neocortex is the youngest of the three, acquiring its more advanced features less than 2 million years ago. How did we come to believe that the youngest is the strongest? Examining neural patterns in the brain challenges the notion that thinking is on top. There are 10 times more signals going from the limbic system to the neocortex than the other way around. Most surprisingly, the brain activity in the neocortex is just 5 per cent of the total brain activity. Thinking comes last by all measurements in the brain's priority for survival. In terms of evolution, this makes a lot of sense.

Even thinking itself is deeply rooted in subconscious brain structures; something that goes almost fully unnoticed in our everyday lives. To show-case this in our seminars and lectures we apply a mini exercise inspired by the neuroscientist Sam Harris (2012). We ask the audience to close their eyes and try not to think anything for as long as they can. After a few minutes we ask them what thought entered their minds first. It is actually the next two questions that surprise them: how exactly this thought was created and by whom. People usually answer aloud: 'Me! I generate my own thought!' Although this seems intuitively true, it is actually an illusion. Most of the thoughts entering our minds come from our subconscious and cannot be altered or controlled by us. These thoughts are significant because, when entering consciousness, they shape our action options for voluntary behav-iour. That is, involuntary (subconsciously driven) thoughts form the basis of our voluntary (consciously driven) actions.

A study by Bhangal and associates (2018) suggests clearly that the contents of our consciousness mind arise involuntarily, without our deliber-ate involvement. The researchers showed participants distinct drawn objects in different colours, and asked them to either name the colours or the shapes, but not both. The fact that half of participants were able to name both the shapes and the colours after the test, regardless of what they had chosen to do (or were instructed to do, as was the case in a variation of the test), with-out trying to do so by choice, supports the existing literature on involuntary brain processes being always present. These processes send thoughts to our conscious mind without us knowing about them or being able to manage them. The percentage of people involuntarily counting objects when they were consciously not trying to do so, in a second experiment by the authors in this study, was even higher than with shapes and colours. As the authors concluded, 'when performing action X, people also experience conscious contents about action Y, even though the latter action was not selected [voluntarily] for production', adding to the extensive scientific literature on the subject. This study highlights the striking fact that the unpredictability of entry of our uncontrolled thoughts which we experience every day indicates the presence of an evolution-based neural system that can be studied and even predicted, and *not* the workings of an un-mechanistic and indeterminate system that is driven by consciousness.

Whenever we present those facts to business executives and others, people are astonished. So deeply rooted is the belief that our brain is a logical, thinking machine that any proof of the opposite creates surprise at best and more often disbelief. This realization has a profound effect on the way leaders take decisions in companies. When it becomes clear that our brains create desired responses to stimuli driven mainly by very old and, mostly, hidden forces, leaders start changing their view on how they make decisions. Their confidence in naturally and voluntarily occurring logic and thinking is shaken. And then they start afresh.

Beware of patterns bearing gifts

So why do we like patterns so much? Employment and customer trends, latest technological developments, macroeconomic cycles, changes in investment and political decision-making, wherever we look there are patterns waiting to be discovered. Patterns that can make or break our business, reinforce or damage our leadership status. We just cannot help it because patterns make us happy. Dopamine, a very influential reward chemical in our brain, is to blame. Whenever we spot a pattern, our brain matches stimuli with memory. This means that whenever a new situation appears, like the latest data on a spreadsheet or new competitors in our market, our brain will desperately search our memory to match this new info with existing previous information. Finding a complete match or creating a new pattern by combining various older information will have the same effect. The brain will be rewarded by a rush of dopamine. No pattern means no explanation. No explanation means more uncertainty. Uncertainty means no survival. In short, patterns create the feeling of control because they increase, realistically or unrealistically, our chances of survival. It is a very effective evolutionary mechanism that pushes our brains to always look for patterns and thus ensuring our continuation as a species.

Dopamine is not to be taken lightly. It is a very strong reward hormone and neurotransmitter in the brain that creates immense pleasure when released. Not all pattern recognition releases large amounts of dopamine, though. Pattern recognition is a normal and automatic brain function that helps us recognize familiar faces, symbols and music. It would be catastrophic if every simple pattern recognition during the day were accompanied by significant amounts of dopamine. Immense pleasure actually accompanies pattern recognition when the pattern arrives in a very difficult and unpredictable situation. This is when we get the greatest pleasure from it. This happens because of the 'prediction error' function of dopamine. Dopamine is strongly released when rewards are least expected and most surprising, not when it is business as usual. Actually, when predictions made by the brain are matched by reality, normal amounts of dopamine are released. The extra amount of dopamine is reserved for the situations that surprisingly exceed expectations. The example from the beginning of the chapter showcases this mechanism in a very clear manner. It was when the situation was critical, the stakes high and no one had a clue of what to do that pattern recognition offered its greatest rewards to the CEO. Alarmed by his initial inability to read the signs, and overwhelmed by the data in front of him, his memory retrieved old patterns to match the situation. When this happened he felt exited, rejuvenated and full of confidence. As Fridman and associates state (2019):

> The brain does not simply react to incoming sensory inputs from the world
> (or from the body); rather it anticipates these inputs by constructing a model
> [predictions].... These predictions are constantly checked against incoming
> sensory input ('prediction error').

If this prediction error is large, meaning that the incoming stimuli do not match greatly the pre-existing model of how reality should look, and this difference is very negative, meaning events in real life punish the brain for its wrong prediction, then the brain learns and updates its internal model in an effort to become more accurate in predicting similar future situations. This means that the dopamine induced predictive system of our brain is a significant learning mechanism to help us survive now and in future situations alike. The problem, though, is that the pain, and thus learning, from detecting wrong outcomes comes when the situation has fully unfolded, which in many business cases is simply too late. So, the *ex ante*, or before the event, excitement and reward of detecting a pattern where there is none often leads leaders to painful lessons *ex post*, or after the event.

This internal model of prediction, which in statistics is called Baysian hierarchical predictive coding, works in the brain in a top-down manner, where curiosity-driven assumptions lead to hypotheses that are tested based on oncoming stimuli information (Dennett, 2018). When we look at financial data, talk to colleagues, interview customers or visit a competitor's website, our brain uses curiosity to catch glimpses of reality and, based on past experiences, to develop a hypothesis of what will happen next:

'Does this mean we will have a cash-flow issue in the third quarter?'

'Is this a sign that our competitors will become more aggressive in their pricing strategies?'

'Did my colleagues have a change of heart concerning our next big investment that I presented last month?'

'Will this innovative technology destroy our market in the next five years?'

Such questions in our minds are usually not as innocent as they seem. They incubate an already developed theory of how things will pan out. Hypotheses are constantly created by our brains without sufficient data from the real world. And the more important the situation, the faster the brain will create its hypotheses, in order to react as fast and as strongly as it can. It is remarkable that, in this process, data streamed from our senses to the brain have less power than the data already existing in the brain. Hypotheses are stronger than reality. Indeed, in visual pathways, there are more top-down than bottom-up pathways: the brain uses more neural avenues to transmit its own ideas (top-down) than it does to challenge those ideas with information from the environment (bottom-up). As Dennett (2018) explains: 'When the organism is on a roll, in deeply familiar territory, the inbound corrections [data from reality] diminish to a trickle and the brain guesses, unchallenged, give it a head start of what is next'.

Executives from around the world explain constantly to us the dynamic nature of their job. Unpredictability is on the daily agenda and complexity is involved in every major decision they have to take. It is in their responsibilities to see the whole picture and offer solutions that will inspire others. At the same time, they need to be as accurate as possible. So pattern recognition is a key aspect of their role in very challenging environments, and a dopamine rush is waiting in every corner. However, leaders with a series of wrong calls eventually lose both their followers and their job. So being aware of this brain trap is more crucial than ever for people in modern organizations.

We cannot avoid seeing patterns in data, people's behaviour and market trends. This is our brain's evolutionary role in ensuring our survival. It is possible to avoid the high of spotting a potential wrong pattern by making sure we examine all possible viewpoints and pay equal attention to diverse opinions. Pattern recognition will emerge from deep inside our brains. Double-checking and cross-checking should be immediate reactions to pattern recognition coming from the top of our brains. Always remember that the gifts of pattern seeking can be great but they can also be the Trojan Horses to your leadership future.

Together we stand... or do we?

'There is no "I" in team', the popular saying goes. The obvious meaning of this statement is that when in a team you have to prioritize the team over you. The hidden meaning, though, is that teamwork can extinguish individual voices for the collective good. But what if these isolated voices were right and the team was wrong? You might recall that in the example at the beginning of the chapter, one of the directors voiced her concerns over the CEO's pattern revelations. But she was immediately shot down from both the CEO himself and his pattern disciples. Finally she went along with them, converting to the team truth. After the disastrous end of the year results she was one of the first to leave the company for another job.

It is not rare in companies to find teams enchanted by an idea, following it as a panacea for all or most of their problems. From a new enterprise resource planning (ERP) system to an expensive rebranding exercise and from a new high-level recruit to an innovative product that will be the definite game-changer, teams fall often into the trap of groupthink. We have encountered groupthink many times in our training and consulting engagements. It usually manifests itself first as a proven solution, then as a strong team commitment before finally revealing its true face – a dogma that no one has the right to challenge. There are a lot of studies, mainly in social psychology, on this phenomenon pioneered by the American psychologist Irving Janis. In his 1982 book titled *Groupthink: Psychological studies of policy decisions and fiascoes*, he views groupthink as an in-group unwilling-ness and blindness to realistically see alternatives to a decision and/or action. Policy, and especially decision-making, groups can become so cohesive and influenced from past success that their leaders can easily gain uncritical support from team members for poor decisions. There are a lot of labora-tory studies that attempted to validate Janis's arguments. Many of them have confirmed that teams with a moderate amount of cohesiveness and more pragmatic collective confidence produce better decisions than teams with low or very high cohesiveness. In contrast, the latter groups made very poor decisions influenced by groupthink.

The neurological aspect of groupthink is that again we witness the inter-play between the deeper brain structures and the executive part. In particu-lar, the domination of the old over the new brain. Driven by an ever-present desire for survival, the brain will push forward agendas that favour group homogeneity rather than challenging the status quo. More so in cases where the beloved idea comes from the charismatic and all-powerful team leader and everyone else seem to enthusiastically go along. Groupthink threatens the individual brain with 'you are either with us or against us', which implies dire consequences for the latter. The choice is ours, but most probably it is has already been made by our brain.

ACTION BOX

In order to avoid or protect your teams from groupthink you can do various things:

- Assign one or two members of the team to critically evaluate decisions all the time. These members can be changed according to the particular case.

- Use different teams to explore the same issue and avoid communication across the teams during this process.

- Avoid involving only experts in the decision-making process. Use people that are less experienced or invite outsiders to be involved sometimes. In that way you can enhance the introduction of fresh and unexplored ideas.

Try the actions above. Also, you need to think of other ways through which you can challenge the status quo of a well-established team, without risking losing team cohesiveness.

Brain misfires

Together with pattern recognition and groupthink, there is a large number of automatic brain responses to situations that make managers take wrong decisions and engage in misguided actions. Those deeper brain structures, the old reptilian brain and the limbic system, having survival as their top priority for millennia, apply fast, unchecked thinking. Being very powerful, they drag our behaviour along with their wishes, giving little chances for the executive brain to interfere strategically. But how deep does the rabbit hole go?

Actually, quite deep. Cognitive biases are now well researched in multiple disciplines (Haselton *et al*, 2005). They occur most often when urgency is heightened, information is confusing and our social surrounding suggests conformity. Our brain's natural limit in information processing causes automated protocols to take over, leading to specific behaviours. Rationality is bypassed. It takes the back seat while the driver is a subjective, closed, impulsive, and self-confirming person. This person cares only about fast responses based on what is already known and comfortable. Risk is unacceptable. Thus, spending energy in exploration and experimentation is seen by the brain as taking away vital energy from more reassuring repeated behaviours. All in all, not a great route for aspiring leaders to follow.

From a neuroscientific perspective, biases are mainly about the ways our brains determine an individual as a member of our group, called the ingroup, or a member of another group, called the outgroup. With more than 1,000 decisions taken per day, and with around 150 biases identified so far, snap judgements of people based on visible characteristics, such as gender (which typically takes about 50 milliseconds to determine) or race (which typically takes about 100 milliseconds to determine) have evolved to help the brain navigate efficiently through a complex and dynamic environment (Jeffery, 2014). Crucial to the success of this navigation is grouping people as friend/ ingroup or foe/outgroup. Reviewing the available scientific literature on the subject, Molenberghs and Louis (2018) systematized the brain processes that are involved in ingroup bias. They suggested five processes that play a key role, as expose in neuroimaging studies:

- *Perception of people.* Neural areas associated with the perception of words, faces and actions are activated when our brain is differentiating people as member of our ingroup or outgroup.

- *Empathy.* Key neural areas associated with experiencing the pain of others are activated strongly when we see ingroup members suffer, while they show less activation for other people.

- *Mentalizing.* Similarly, core neural areas associated with thinking about other people's mindset are more or less activated depending on which outgroup people belong to. For some outgroups, where the perceived threat is high for us, it activates strongly while for outgroups we do not care about it activates weakly.

- *Moral sensitivity.* Our brain is much more sensitive when confronted with people from an outgroup than when dealing with people in our ingroup.

- *Reward system.* The brain's reward mechanisms activate more strongly when we socialize with people in our ingroup, but also when people in the outgroup suffer.

Brain-based leaders would use the above as criteria to identify ingroup and outgroup attitudes within their teams. If ingroup mentality and discriminatory behaviour are based on basic human characteristics such as gender or race, leadership response should be decisive and timely. Such phenomena should not be tolerated under any circumstances, both for humanistic and business reasons. Recognizing this necessity, many companies the world over have engaged in training schemes that aim to minimize unconscious biases against people in the workplace. However, such training schemes have not brought the desirable effects. The *Diversity Management That Works* research report by CIPD, the professional body for human resources and people development in the UK, found that such training did not produce any conclusive behavioural and attitudinal improvements (CIPD, 2019). Training did manage to increase awareness of these issues, but this awareness did not necessarily translate into meaningful change. Brain science could predict this, since understanding, or cognitive processing, does not predictably lead to a new behaviour. As training fails to induce behavioural change, the same goes for information. Maxine Williams, Google's Global Director of Diversity, emphasizes that, in changing attitudes and behaviours, 'numbers can take us only so far' and that 'algorithms and statistics do not capture what it feels like' to be discriminated against and to experience a biased corporate culture on a daily basis (Williams, 2017). According to Williams, a deeper and more personal approach is needed. This means that

changing people's minds through training and information are not sufficient, or even prerequisites, for changing their behaviour. This book offers a
variety of scientifically backed ideas and practical tools for achieving change
by utilizing diverse brain functions, not just understanding.

In particular, this book, beyond making you more aware about these
issues, supports the idea that being more resilient to biases depends on to
what extent you can initiate yourself to do so. One idea that we use extensively in our lectures and coaching seminars is suggested by Howard J Ross
in his book *Everyday Bias* (2014). His argument is simple, but critical for
everybody and especially for leaders. He claims that in a (business) culture
full of values of certainty, it is extremely important to identify the ways
that alert us and make us more consciously uncertain. How can we do that?
By simply taking a PAUSE and rethinking what we intend to do about
something. For Ross, PAUSE as the key to overcome biases is based on the
following aspects:

P – Pay attention to what's happening beneath the judgement (event vs
interpretation).

A – Acknowledge or identify your reaction/interpretation/judgements.

U – Understand other possible reactions/interpretations/judgements.

S – Search for the most constructive/empowering/productive way to deal
with the situation.

E – Execute your action plan (act consistently with what makes the most
sense).

Put it in more simple terms, this means that, as leaders, you need to think
twice. Before you take a critical decision like hiring someone, firing someone, evaluating and rewarding someone, approving an investment, even calling a meeting, think twice about the reason for your decision. Give yourself
the benefit of the doubt. Our own experience has proved this is a very practical as well as effective and relatively fast way to respond to possible biases.

If group biases are based on other, more organizational-centred criteria,
such as internal competition, departmental conflict, power tribes and
personal ambitions then the brain-based leader must first recognize them,
then assess their impact, and then enable the formation of appropriate group
mentalities that boost team and company performance. Famously, Steve
Jobs, the iconic leader of Apple and Pixar, created in Apple an internal
culture of extreme antagonism with other IT giants, most notably Microsoft.
So much so, that he pursued with heightened passion the launch of the first

iPhone because, as reported by the former head of iOS, Scott Forstall, he met a Microsoft executive who bragged about Microsoft's work on tablets at a social event (Gibbs, 2017). Creating a sense of 'us' against 'them' is not inherently always bad and leaders can harness the power of our brain's natural aptitude for group perception by creating the right ingroup vs outgroup attitudes, both in teams and in the organization as a whole.

The threat of cognitive biases, especially in ingroup vs outgroup relations, is ever-present and can be expressed in any managerial situation. When you label people and apply stereotypes, when you are not open to new and conflicting information, when you trust your memory and experience too much, and when you place yourself in the middle of the universe, cognitive biases are at work (Ariely, 2008; Kahneman, 2011). The brain does not direct enough energy to the prefrontal lobe for a more objective analysis of the situation. It prefers to resort to less energy-demanding neural patterns and see things as they should be and not as they really are. And as we have discovered through our engagement with hundreds of companies in more than 20 countries, our modern business environment cultivates corporate cultures that push for biases to surface more often than not.

Corporate cultures favouring the biased mind

A social psychology study on a biblical story reveals how companies build cultures that favour biases. In 1973, Darley and Batson published their now classic study on the parable of the Good Samaritan. The parable is about someone helping a person in great need without pursuing any selfish goals. The two scientists ran an experiment at Princeton asking their students to come to a building to talk about the Good Samaritan and another unrelated topic. Then they were instructed to run to another building for an exam. Some of the students were asked to hurry to get to the other building while others were asked to hurry less. On the way to the other building, all students passed by a person asking for immediate help and looking like he really needed it. Who helped more? Surprisingly, talking about the Good Samaritan story prior to encountering the person in need had no effect on students' reaction. What had a real effect on their behaviour was urgency. Only 10 per cent of students who were asked to hurry to the other building stopped to offer help, while more than 60 per cent of the students who were not in such a hurry stopped to help.

Many companies globally have corporate values telling a similar story. Customer-centric, solutions-oriented, team-supporting, employee-promoting, community-focused are values we constantly encounter in corporate materials around the world. What we actually see inside companies, though, is that urgency cancels those values, prioritizing self-centred, short-term goals. A tight deadline, the next big project starting now, the next morning meeting you need to prepare yourself today for, the latest numbers that were not as good as expected, the new director that wants more and more, and other emergency calls create cultures that nurture automatic behaviours, frontal lobe bypassing and cognitive biases. Regardless of the nice narratives of good deeds in businesses, the pressure of 'now' decreases the brain's ability to direct energy to vital thinking structures that can ask the right questions. Instead of asking, we are just doing. Most importantly, we fail to stop and ask the man in need 'How can I help?' or 'How can I make the situation better?' as in the Princeton experiment. In companies today, questions like these addressed to our teammates, subordinates, bosses, people in other departments, and of course to ourselves are crucial to keep biases and knee-jerk reactions at bay. But to do this, we need to start asking and stop rushing.

A study on how arousal changes behavioural intentions is also very helpful in supporting corporate cultures that are more considerate and reflective. Ariely and Lowenstein (2006) asked students to answer questions on moral behaviour, instructing them to answer the same questionnaire later when in a state of arousal. The answers were significantly different, showcasing graphically that well-considered intentions can go out of the window when specific brain sections get aroused. We know to do the right thing when calm, but when aroused we behave differently. The study focused on sexual arousal but results are not confined to this brain activity exclusively. Arousal involves the activation of our automatic nervous system and leads to increased heart rate, higher blood pressure and an extreme sensitivity to external stimuli. In such cases, rational thinking loses control and power lies with more impulsive and pre-programmed reactions.

The case of Uber, the globally famous ride-hailing company, is a prime example of how an overly strong focus on fast growth can lead to an aggressive internal culture, overriding any good intentions and carefully drafted corporate values. As was widely reported in early 2017, the company's culture was found to include systemic instances of discrimination, harassment and hostile behaviour (Isaac, 2017). Regardless of the CEO and founder Travis Kalanick's extensive efforts to fix the problem, he had eventually to step down from his CEO position a few months later (Kleinman, 2017). This example illustrates emphatically that, regardless of the cognitive safeguards we might put in place to make sure that bad behaviour will not manifest in the workplace, our daily behaviours and inner passions can jeopardize everything by prioritizing the same behaviours we verbally reject, signalling to everyone around us that this is what they need to do to survive in this company.

Our corporate narrative might be promoting all the great values of understanding, collaboration and reflection, but urgency and the aroused/stressed state of mind we constantly find ourselves in at work makes us more subservient to our biases than ever. The answer is questioning more.

Asking is winning

The most reliable managerial weapon against cognitive biases is actually the simplest one: asking questions. A lot of them, and all the time. Asking the right questions, and even some wrong ones, creates the appropriate conditions for challenging automatic reactions and thus revealing any biases that might be surfacing. Questioning is about getting to a mental state of observation prior to action. Inserting this intermediary step of questioning minimizes the immediate behavioural effect of biases. Instead of the brain jumping directly from a stimulus to a premature conclusion or to a misguided behavioural reaction, a series of questions has valuable stopping-power. That is, it has the power to interrupt fast neural processes and impulsive activity, re-directing more brain energy to thinking and reflecting functions.

The power of asking questions in companies and other institutions in our societies is finally being noticed. According to Warren Berger (2014), a major advocate of informed enquiring, our brain is hungry for information from the time we are born, with 40,000 questions asked between years two and five and 300 questions asked daily at the age of four. However, the rate of questioning peaks at five years old and declines soon after. The reason for this is the anti-enquiring culture we have built at home, in schools and at work. Instead of encouraging questions as one of the most natural ways for our brain to learn and change behaviour, we have neutralized it. When you are faced with the attitude of 'it is what it is' and 'this is how we do things here' you instantly know you are in an environment that is hostile to informed enquiry. But asking questions is just the beginning. Asking the right questions is of equal importance, too. Schoemaker and Krupp (2015), based on their extensive corporate research, have categorized key questions asked by winning leaders into six groups:

1 Questions that make you think outside the box and uncover the hidden meaning of wider market trends.

2 Questions that help you explore future scenarios and analyse major external uncertainties for each one.

3 Questions that help you become a contrarian that examines every problem from multiple and diverse angles.

4 Questions that help you identify the right patterns by deploying multiple lenses for discovering difficult-to-see connections.

5 Questions that help you create new options by evaluating numerous alternatives and their unintended consequences.

6 Questions that help you learn from failure fast and use this as a source of immediate improvements and further innovations.

For them, the starting point to long-term leadership success is not in answers but questions. The list above is a very effective remedy for battling cognitive biases and limiting the brain's obsession with the known.

ACTION BOX

By using the framework above, try to develop some questions for each of these six types. Try to think of specific situations in order to develop these questions. Ask some of your people/team members to do the same. Can you spot the difference?

Socrates famously said, 'I know one thing, that I know nothing'. His questions were so powerful that in the end they earned him capital punishment. Societies with a slow pace of change, as the ancient ones were, do not like challenges to the status quo. Companies of the previous paradigm, operating in stable industries, did not really foster a culture of enquiry. Today things are very different. Examples of CEOs turning around companies by the power of questioning, supporting and learning are now common. Frank Blake, the CEO of Home Depot in the US, who guided the turnaround of the company in the very challenging times after the 2008 crisis, did so by knowing very little about the company and the industry when he took over the CEO position (Reingold, 2014). His commitment to healthy and productive enquiry at all levels and his recommendation for learning walkabouts in stores for his key managerial personnel lead to the right changes that increased customer satisfaction and brought higher share value.

However, there is another set of questions that modern leaders need to ask. This set concerns their most strategic challenges: the ones that deciding upon can have significant impact, positive or negative, on the lives of people inside and outside the company. These questions are less about the business itself and more about the wider role of the company in the world and its responsibility as an active shareholder in society. These tough decisions require a specific enquiring mindset that puts to the test assumptions, habits and seemingly unavoidable strategic necessities. These are the questions that make leaders achieve greatness beyond their immediate role as individuals who steer businesses. According to Harvard Professor of Business Ethics, Joseph Badaracco, these questions are (2016):

- *What are the net consequences of your decision?* This question entails rigorous analysis of possible courses of actions together with their human impact. Rushing into decisions, especially ones with wider effects on the lives of diverse internal and external communities, is the wrong approach. Although the future is largely unpredictable, putting together the right team, spending time on the matter, and examining human costs and benefits of possible decisions are the minimum requirements for answering this question successfully.

- *Which are your core obligations in this decision?* Apart from the leader's obligations to shareholders and to their teams, there are other obligations that go beyond the strictly business-centred scope of work. Such obligations are pan-human, since they relate to all possible humans potentially affected by a decision. For example, the obligation to protect people's dignity as human beings or the obligation of basic human rights cut through short-term business objectives and, if performed appropriately, can cement a leader's position as the guardian of both the company's and humanity's future.

- *Do you see the world as it is?* This question includes two subtopics: do you see the outside world as it is, and do you see the inside world as it is? The first has to do with the practicalities of implementing your decision, avoiding any illusions and misconceptions about the intentions and capabilities of others. The second concerns your capacities, readiness and willingness to undergo the needed changes, challenges and pathways to implement the decision.

- *Which are your communities?* Since our brain is a social organ, probing and recognizing the diverse communities you belong to are of paramount important in taking the right decision. This question requires you to identify the values and norms that guide your corporate behaviour. At the same time, it asks you to challenge those values and norms in order to make sure that you are not taking decisions merely to preserve existing ways of thinking and doing.

- *Will you live with your decision?* In the end, after any decision we make, we need to live with it and take full responsibility of its consequences, positive or negative. In order to explore if the decision you want to make corresponds with your inner compass, you need to take time to reflect, speak to mentors or professionals you trust and respect the most, and confide in those that know you personally the best.

The power of asking in action brings more energy to the rational part of your brain and minimizes the impact of neural shortcuts. But what kind of an enquiring leader do you want to be exactly?

Enquiring with style

Asking the right questions is still not enough. We also need to enquire with the right attitude in order to make sure that the outcome will be beneficial for all. We have experienced cases where directors practised the art of questioning aggressively and in a non-productive way. Many times, this type of questioning creates an extremely uncomfortable environment where questioning looks more like interrogation than collaborative thinking. Table 2.1 shows our questioning style matrix.

TABLE 2.1   Questioning style matrix

| | | Engagement | |
		Active	Passive
Attitude	Constructive	The star enquirer	The good listener
	Non-constructive	The evil interrogator	The grumpy loner

We have identified four types of questioner:

- *The star enquirer.* The leader is actively engaging in conversations, many times as the initiator and facilitator. All discussions are friendly and often creatively heated, and there is no wrong answer. The aim is mutual development.

- *The good listener.* The leader is always open to answer questions and provide feedback. However, it is usually other people who initiate discussions. Listening, guiding and deciding are main characteristics, but in a reactive manner.

- *The evil interrogator.* The leader asks everyone difficult to answer and even offensive questions. The main purpose is to make others feel inferior and incapable. This leader is feared, not respected.

- *The grumpy loner.* The leader is isolated by their own choice, giving every signal possible to others to keep away. Makes people feel uncomfortable when disturbed and looks like everything is under firm control and should not be questioned. Questioning is associated with weakness.

Moving towards the star enquirer quadrant is essential for a leadership-calibrated brain. It might not be easy to change, but it is necessary in order to fight biases and other mental shortcuts.

ACTION BOX

By using the framework above, try to evaluate yourself. Think about each of these types and try to classify yourself in one of the categories. Write down what you do that puts you in that category. Think what you need to do in order to be able to move categories. Write down your thoughts again. It is important to write down your thoughts since this can help you develop a much more structured and concrete thinking process around these issues.

LEARNING FROM REAL LIFE: THE TALE OF TWO MANAGERS

Manager A asks the right questions, at the right time, of the right people.

Manager B never asks, in order not to appear weak.

Manager A encourages independent thinking.

Manager B allows you to express any opinion in meetings as far as it confirms the manager's own.

Manager A always double-checks data... and then checks once more.

Manager B takes one look at data and moves straight to implementation because the pattern is always evident.

Manager A has a 'to the best of my knowledge' attitude.

Manger B has a 'my knowledge is the best' attitude.

Manager A exposes themself to the ideas of outsiders.

Manager B sees the views of outsiders as irrelevant.

Manager A forces the team to individually and collectively reflect often on decisions and projects.

Manager B has no time for reflection, only time to start another project.

Manager A favours research that involves multiple samples and a variety of data collection methods.

Manager B has used the same research approach for years because it has been tested and works.

Manager A attends conferences, industrial meetings and training with a productive, open-minded attitude.

Manager B always comes back from an event with the phrase 'I knew that already, nothing new there.'

Manager A follows top thinkers in diverse fields on Twitter, blogs and in publications.

Manager B does not follow anyone (at least openly) – others should follow him.

Manager A allows themself regular breaks during the day and periodic longer escapes to re-set their mind.

Manager B believes that breaks are for the weak.

Manager A takes an interest when they spot a bias in themself or in others, being always polite and gentle about it to themself or to others.

Manager B never accepts that an error of thinking came from their own bias because their thinking is based on pure rationality. The manager thinks everyone else is full of biases.

Manager A considers every opinion carefully (though not necessarily for the same length of time) even when it comes from a chance conversation outside the office with people not related to work.

Manager B never talks business to anyone outside the office, except to boast about a success.

Manager A tests and uses the latest apps for productivity, collaboration and idea-generation, and loves experimenting.

Manager B uses emails and spreadsheets, and hates experimenting, believing that this shows that people don't know what they are doing.

Unfortunately, we have yet to meet a person that stands exactly in between these two extremes. Although it is not necessary for someone to possess all characteristics of manager A or B we constantly find that most characteristics will be in one or the other category. Those two managerial types correspond with Philp Tetlock's (2005) classic categorization of political forecasters of foxes and hedgehogs. In his study *Expert Political Judgment: How good is it? How can we know?* he explores what makes people predict better political outcomes. He recalls an ancient Greek poet that famously said that foxes know many things while hedgehogs know only one great thing. In Tetlock's widely accepted view (Silver, 2012), people's general attitudes critically affect their thinking and prediction capabilities. A fox is open-minded, accepts responsibility for mistakes, always learns and sees the world as a complex and difficult to predict system. The hedgehog is stubborn, always seeking to confirm old beliefs, does not listen to outsiders and is happy to explain everything according to existing theories. Foxes, on aggregate, are far better forecasters than hedgehogs. Manager A is a fox with a brain that is better calibrated to become a great leader in dynamic and chaotic times. Manager B is a relic waiting for extinction if they don't change fast.

When the ape takes control

One of the most significant lessons we learned early on in our quest for understanding the brain's impact on leadership, and on human behaviour in general, was the existence of significant survival protocols. The brain has its own ancient protocols deeply rooted into its hidden neural networks that are activated in extreme cases to ensure our survival. And we learned it the hard way.

One of us had a car accident at the end of the 1990s. Nobody was seriously injured in the incident, but remarkably enough the exact moment of the crash was not experienced consciously because the person blacked out.

Regaining consciousness a few seconds after the impact was crucial for taking the necessary steps to escape with minimum injuries. The big question, though, was who decided to shut consciousness down at the exact moment of the impact? Surely, this was not a rational decision, since everything happened in milliseconds. Searching for an answer led to the realization that the brain applies its own rules when it decides that a situation is critical. External stimuli travel through our senses to the brain, which, faster than thinking, decides how to treat them. In the case of the accident, the brain decided to blackout perception in order to avoid clogging the system with a shock. If in shock, then behaviour might not have been optimized for survival (Goleman, 1998). We can call such protocols *hijacking*. But is the hijacking of executive brain functions from more primitive ones happening only on extreme cases? No. It happens every day. Modern leaders need to be particularly aware of a brain hijacking mechanism known as amygdala hijacking that can threaten their clear thinking on a daily basis.

The amygdala is part of the limbic system of the brain, which, as described earlier, is the system that deals more with memory, learning, motivation and emotion. Within this system, the amygdala plays the role of emotional guardian of survival, deciding if and when to interfere. Interference means taking over. The amygdala is generally associated with emotional learning since it matches external events with internal memory to choose the best possible action. If the external event is threatening to evoke a strong, unpleasant memory then it decides to take control by stopping rational lobes from over-thinking. In a situation of severe danger there is no much time for debating and deliberating. Consequently, the amygdala is also associated with stress, anxiety, fear and aggression (Goleman, 1998). For leaders, this means that there are daily situations at work that the amygdala can consider as threatening, and will hijack higher thinking, with important consequences. Those consequences include erratic behaviour, mistrust of others, low self-confidence, impaired decision-making and performance, inability to read signals (human and numerical), self-centeredness, short-term thinking, misunderstandings and miscommunication, knee-jerk reactions and revengeful attitudes. Hardly the desired characteristics of inspiring individuals.

However, we are not defenceless against amygdala hijackings. There is a lot that can be done. Professor Steve Peters (2012) from the University of Sheffield has developed a mind management programme by which he helps athletes and other high achievers to deal effectively with such hijackings. According to his approach, our internal 'chimpanzee' who, courtesy of evolution, is five time stronger than our conscious will, influences our behaviour by attacking our mind with poisonous and fearful thoughts like 'I can't do this' and 'Everyone is against me.' Typical statements that give away the internal chimp's intentions are:

'Things will not go as I would like them to' (external issues).

'I do not feel fit for the task in front of me' (internal issues).

The way to deal with the ape that lives in our head and has the power to override our clear thinking is first to accept it and avoid fighting with it (it is stronger). We need to observe all its moves and impulses and then to nurture its better side by making it feel less insecure in the long-term. Finally, we can manage the chimp through a system of distractions and rewards. A distraction can be something that will not allow the chimp to express themself fully, like quickly entering the presentation room without much last-minute stressing about it. A reward is about giving the chimp something it likes when the work it is opposing to has been done, partially or totally. This could be something like having your first coffee break only after you reviewed the difficult report in the morning.

A more physical approach in dealing with hijacks is advocated by Dr Alan Watkins in his book *Coherence: The secret science of brilliant leadership*. Dr Watkins (2013) suggests a relaxation method through breathing that can calm the irregular heartbeat caused by a mind full of amygdala-produced fear. Taking the time to breathe regularly, smoothly and through the heart on a daily basis can have a significant impact on stress reactions. It can grant the mind the right space to develop constructive feelings and thoughts. Ignoring the physiology of hijacks is a strategic mistake that reduces the ability to deal with them successfully. We always recommend to our clients to take a short break from a situation when they feel overwhelmed by anger or fear. The popular advice of never sending an email when angry, but rather leaving it for the next day, has its origins in the interdependence of physiology and psychology. Take a walk, breathe calmly and regularly from your chest, and allow your system the necessary time to flush out chemicals produced by aggression and fear. Those simple acts can do more than you think for boosting your leadership performance.

The key issue of managing hijacks is to be aware of them, and separate them from yourself. If you treat such negative thoughts with a genuine interest rather than with a fatalistic acceptance, half of the job is done. Your thoughts are not necessarily you. Although you cannot easily decide what comes into your mind, you can decide what to do with those thoughts, especially when your body is in a calmer state. It should be your clear mind leading your people, not your internal ape.

Keep in mind

Our mind can never be free from irrational influences pushed up by the older brain structures. However, by resisting the dopamine high of pattern recognition and the gravitational pull of cognitive biases, and by being aware of the internal ape that wants to get the best of us, we can achieve a clearer state of mind. Constantly being aware of biases and asking the right questions, and creating a culture that values debating, will increase the strategic role of clear thinking in our daily leadership challenges. This in turn will allow us to give clearer and more meaningful directions to our team.

BOOST YOUR BRAIN: YOU AND YOUR APE

Think about the ape side of your brain. Make a list of things that you have done when the ape was in control. What were those situations about, and can you categorize them? Then make another list with potential actions as a leader through which you need to control the ape inside you. Consider your conclusions and be aware of them next time one of those situations is repeated.

References

Ariely, D (2008) *Predictably Irrational: The hidden forces that shape our decisions*, HarperCollins, London

Ariely, D and Loewenstein, G (2006) The heat of the moment: The effect of sexual arousal on sexual decision making, *Journal of Behavioral Decision Making*, 19(2), pp 87–98

Badaracco, JL (2016) How to tackle your toughest decisions, *Harvard Business Review*, September, pp 104–07

Berger, W (2014) *A More Beautiful Question: The power of inquiry to spark breakthrough ideas*, Bloomsbury, New York

Bhangal, S, Merrick, C, Cho, H and Morsella, E (2018) Involuntary entry into consciousness from the activation of sets: Object counting and color naming, *Frontiers in Psychology*, **9**, p 1017

CIPD (2019) *Diversity Management That Works: Research report*, www.cipd.co.uk/ Images/7926-diversity-and-inclusion-report-revised_tcm18-65334.pdf (archived at https://perma.cc/5VYQ-EBL4)

Darley, JM and Batson, CD (1973) 'From Jerusalem to Jericho': A study of situational and dispositional variables in helping behaviour, *Journal of Personality and Social Psychology*, **27**(1), pp 100–08

Dennett, DC (2018) *From Bacteria to Bach and Back: The evolution of minds*, Penguin Books, London

Fridman, J, Barrett, LF, Wormwood, JB and Quigley, KS (2019) Applying the theory of constructed emotion to police decision making, *Frontiers in Psychology*, **10**, p 1946

Gibbs, S (2017) The iPhone only exists because Steve Jobs 'hated this guy at Microsoft', www.theguardian.com/technology/2017/jun/21/apple-iphone-steve-jobs-hated-guy-microsoft-says-scott-forstall (archived at https://perma.cc/ V5TK-XAK4)

Goleman, D (1986) *Emotional Intelligence: Why it can matter more than IQ*, Bloomsbury, New York

Goleman, D (1998) *Vital Lies, Simple Truths: The psychology of self deception*, Bloomsbury, New York

Harris, S (2012) *Free Will*, Free Press, New York

Haselton, MG, Nettle, D and Andrews, PW (2005) The evolution of cognitive bias, in *The Handbook of Evolutionary Psychology*, ed DM Buss, pp 724–46, John Wiley and Sons Inc, Hoboken

Hill, D (2010) *Emotionomics: Leveraging emotions for business success*, Kogan Page, London

Isaac, M (2017) Inside Uber's aggressive, unrestrained workplace culture, www. nytimes.com/2017/02/22/technology/uber-workplace-culture.html (archived at https://perma.cc/2USF-8QNE)

Janis, IL (1982) *Groupthink: Psychological studies of policy decisions and fiascoes*, Houghton Mifflin, Boston

Jeffery, R (2014) The neuroscience of bias: Diet cola makes you (even) more racist, www.peoplemanagement.co.uk/long-reads/articles/neuroscience-bias (archived at https://perma.cc/A8P7-R2TK)

Kahneman, D (2011) *Thinking, Fast and Slow*, Farrar, Straus and Giroux, New York

Kleinman, Z (2017) Uber: The scandals that drove Travis Kalanick out, www.bbc.com/ news/technology-40352868 (archived at https://perma.cc/L6E2-6NMG)

Molenberghs, P and Louis, WR (2018) Insights from FMRI studies into ingroup bias, *Frontiers in Psychology*, 9, p 1868

Peters, S (2012) *The Chimp Paradox: The mind management program to help you achieve success, confidence, and happiness*, Vermilion, London

Reingold, J (2014) How Home Depot CEO Frank Blake kept his legacy from being hacked, *Fortune*, 17 November

Ross, HJ (2014) *Everyday Bias*, Rowman and Littlefield Publishers, London

Schoemaker, PJ, and Krupp, S (2015) The power of asking pivotal questions, *MIT Sloan Management Review*, 56(2), pp 39–47

Silver, N (2012) *The Signal and the Noise: Why so many predictions fail-but some don't*, The Penguin Press, New York

Tetlock, PE (2005) *Expert Political Judgment: How good is it? How can we know?* Princeton University Press, Princeton

Watkins, A (2013) *Coherence: The secret science of brilliant leadership*, Kogan Page, London

Williams, M (2017) Numbers take us only so far, *Harvard Business Review*, November–December, pp 142–46

03

Higher performance, more followers

IT'S ALL FOR THE SALES... OR IS IT?

She gladly accepts the new position and the responsibility of running the global sales operations. The CEO has made it clear that he trusts her fully. He has also stressed the fact that she is to develop the new sales strategy that will bring them in par with other global players. The future of the company is in her hands and she is fine with that. Throughout her life, whenever she had to deliver results, she never felt scared. Her confidence was based on her ability to prioritize correctly, roll up her sleeves and tackle the problem head-on. And it always worked. But would it work now, for the most important role in her career?

It starts well. She develops the main strategic direction herself, with key inputs from her peers and a couple of seasoned consultants. The strategy is very well thought out, with great analysis, and is already accepted with praise by the board. It is a bold plan, including innovative approaches. Times are challenging and require fresh thinking and deep changes to solve new problems. What is left is to present the plan to the regional sales directors in order to start implementing it globally as soon as possible. She has done big and important presentations before. Nevertheless, this is *the* most important presentation in her career. Naturally, she stays up at night perfecting every sentence and carefully designing every slide. She rehearses and then she rehearses again. Before entering the hall she runs through the numbers once more in her mind and makes sure she has everything in the right order. She is aware of the fact that such changes are a tough sell, but she also knows that every single word and number in the presentation are there to help the company survive. And they all want the company to do well, right? If not that, then why does she work so hard? If it's not for the sake of more sales then what is it for?

However, despite the highly polished speech and visuals, the clarity of the numbers and the conviction of her voice and gestures, the presentation failed to impress. Sure enough, everyone got the point and they agreed on its usefulness. But something significant was missing. The main comments afterwards, mostly during the informal dinner, were that the strategy is spotless but that the necessary changes do not feel engaging. Even the CEO, talking to her briefly at the dinner, remarked that the room was rather cold during and after the presentation. Next morning, she did what she always has in such situations: rolled up her sleeves even more and started working harder to make the plan come true. Implementation of the first phase shows regional directors performing well but not as well as she would like. Some are doing better than others, but overall they are not as supercharged in changing their ways as she would expect them to be in such challenging times. In all honesty, even she is not as enthusiastic as she imagined she would be in the beginning of the assignment. Something is missing for her as well. Learning and change are slow and painful for all. After a few months, she steps down. She no longer has the inner drive to get her going and to over-perform as in the past.

What happened? The task was clear, numbers were strong and everything was in order for success. Everything apart from her brain.

Let's be honest: great leadership goes hand-in-hand with great performance. Our ability to perform with excellence and to influence others to do the same depends not only on our traditional view of performance, namely on a mix of skills, competencies, time, resources and guidance. It also depends on the brain's ability to recognize the significance of the task in hand and to produce the commitment, endurance and changes needed for success, regardless of barriers, failures and shortcomings. This ability comes from a remarkable brain characteristic that surprises everyone coming across it for the first time. Apparently, our brain can change a lot during our lives and this change can determine the outcome of our efforts in a critical and decisive way.

The ever-changing brain

The brain never stays the same. It constantly changes. The brain is a dynamic system of neurons that are constantly moving. This remarkable fact has taken neuroscience by storm and disproved long-held beliefs that all brain changes happen only when we are very young. As reported by psychologist and neuroscientist Elaine Fox (2013) in her book titled *Rainy Brain, Sunny Brain: How to retrain your brain to overcome pessimism and achieve a more positive outlook*, it used to be the mainstream notion up to the 1980s that the brain is a biologically static organ, meaning that its structure and nature do not change inherently during our adult lives. So, more or less, we got stuck with our brain for our whole life. Our duties were, first, to make sure we did not damage it in any way and second, to use it as best we could. Although it could always get worse, through accident or illness, it could never get better. The upper limits of our brain were set in stone since we could not drastically improve its learning abilities and performance. This is why IQ tests were so important. We had a stable brain capacity that we could measure and work out its potential. This cannot be further from the truth.

For more than two decades now we have known that the brain is anything but stable. Our brain is plastic, meaning that it has the ability to rewire itself, creating new synapses between its neurons based on what is happening inside and outside of us. Our understanding of neuroplasticity, as this ability is called, has altered the way in which we view the brain and our capacities in life. The way we use our brain can actually change it, negating the previous view that the matter in our brains is the rigid 'hardware' and our thoughts are simply the 'software'. Neuroplasticity shows that our 'software' can change the 'hardware', although this is only a rough analogy since in order to improve our smart devices we need to purchase hardware upgrades and technical plug-ins. Instead, in our brain the 'hardware' upgrade can happen by upgrading the 'software', or by changing our behaviour. As put nicely by Chopra and Tanzi (2013) in their book *Super Brain: Unleash the explosive power of your mind*:

> Neuroplasticity is better than mind over matter. It's mind turning into matter as your thoughts create new neural growth… Our brains are incredibly resilient; the marvelous process of neuroplasticity gives you the capability, in your thoughts, feelings, and actions, to develop in any direction you choose.

The remarkable ability of our brain to rewire connections between its neurons has led to the now famous phrase 'cells that fire together, wire together', which actually suggests that the more we use particular neural pathways, the more these pathways will be strengthened. Although the advantages of such neural activity are numerous, it also has its downsides. First, connections between neurons can become too strong, leaving little room for flexibility and change, and second, areas that are not used can grow very weak. 'Use it or lose it' is another famous phrase on neuroplasticity. It means that if you do not adapt and evolve new ways, you do not just remain the same as before; you actually deteriorate your neurons and their connections, becoming less able to perform some tasks than before. This is particularly true for cognitive rather than motor skills: we cannot really forget how to ride a bicycle, but we find it very difficult to speak a language we learned years ago and did not practise for a long time. Since most managerial and leadership skills in modern companies are related to cognition and not to manual work, this means you need to be very careful in retaining strong neural connections where they matter the most. Otherwise, you become cognitively weaker and professionally more vulnerable.

Practically speaking, though, how you can do that? How you can develop, grow and maintain your leadership cognitive skills? Well, one way is to read books (like ours) and participate in leadership development programmes. However, do not expect much. You can definitely read about nice ideas related to leadership capability, and have fun and acquire some useful knowledge at leadership training programmes, but developing your actual leadership skills is quite a different story, we are afraid. Leadership as a cognitive capability can be developed and maintained mainly by doing it. You need to let yourself 'enter' a real leadership situation. You need to actually think and act like a leader in order to see improvements in your leadership skills. And not just once. You need to do it more and more, since your 'leadership-related' brain cells need to be always fired in order to be always wired and active.

In an unpredictable and continuously changing environment where organizations need to keep up with changes coming from everywhere, leaders' brains need to constantly adapt and evolve. Neuroplasticity can be our best friend or worst enemy, depending on how we use our brain. Neuroplasticity works to a leader's advantage through the process of constant learning, since learning is the main fuel for neurons to create new connections and paths. Bruce Hood (2014), a renowned British psychologist, clearly stated in his book *The Domesticated Brain* that there is positive plasticity in our brains, even in our adulthood, as far as we constantly learn throughout our lives. Learning is the key to plasticity as it is the key to performance for modern leaders. As Indra Nooyi, Chairman and CEO of Pepsico, advised professionals in *Fortune* magazine (2014):

> Never stop learning. Whether you're an entry-level employee or a CEO, you don't know it all. Admitting this is not a sign of weakness. The strongest leaders are those who are lifelong students.

A plastic brain that never stops learning is a brain with improved cognitive skills such as better memory and enhanced attention. So, it plays an instrumental role in developing our neocortex capacities, and this is the main reason it appears within the first pillar of our model. However, it is not a neural superpower destined for perpetual use. It declines over time, and it can decline faster with our mindless help. We are responsible for our neuroplasticity, and this is one of the strongest messages that modern neuroscience has to offer. As Adam Gazzaley, a neuroscientist at the University of California, explains (Medeiros, 2017):

> Like every organ in our bodies, the brain undergoes changes in how it performs. You see it in your muscles, your bones, your hair – and you feel it in your brain. That is not helped by people seeking comfort and a less demanding life when they are older. The fact is that the brain is still plastic even when they are 70 or 80 years old. It can still be optimized – but instead, many people unwittingly accelerate its deterioration.

But how do we keep ourselves motivated to constantly learn, adapt and evolve? Where do we find the inspiration and the drive to go first where no one has been before and to take others with us? Most importantly, how do we keep our brain alive, firing and growing through challenges and even through disappointments? First of all, a leader with an active plastic brain is a leader with a strong purpose.

Purpose above all

The main problem with the new global sales director in the opening case was that she did not have the right attitude in her new role. Although she managed the tangible elements of her job well, like numbers, strategies and presentations, her performance was not driven by a higher purpose. And the higher you climb the organizational ladder, the more you need a great purpose to drive you, guide you and make those around you follow you. This is something that no modern leader can ignore anymore. Instead of focusing on sales as the ultimate goal and main motivator both for herself and the others, she should have targeted different purposes, activating other brain regions.

In one of the most-watched TED talks online, Simon Sinek (2009) explained the simple biology of purpose. Starting from inside the brain and using the famous cases of Apple, the Wright brothers and Martin Luther King, he famously argued that the most powerful motivation to do something comes from deep inside the brain. These older brain structures are moved by passion deriving from answering convincingly 'why' questions. In contrast, most companies and people try their best to answer to 'what' and 'how' questions, forgetting that the true power lies not in the complicated explanations of products, process and procedures but in big ideals. A great and preferably long-term purpose can help people around you to see the direction and understand better the decisions made and the decisions that will be made. Sinek (2010) in his book *Start with Why: How great leaders inspire everyone to take action* described his golden circle that holds in its centre the mighty 'why' question. If you are driven by a powerful 'why', which is related to a higher purpose, then you will put in the hours, commitment and even the influence needed to achieve your goals. If you are driven by 'what' and 'how' questions, you will achieve very little in difficult times and bring with you nobody. 'Why' brings passion and connections from deep inside the brain, the areas that really drive behaviour, while 'what' brings rationalization, analysis and very little movement. This claim is remarkably similar to Daniel Pink's assertion, in his book *Drive: The surprising truth about what motivates us* (2009a) and in his own famous TED talk on the puzzle of motivation (2009b), that contemporary problems require creative problem solving that can only come from intrinsic motivation. If you chase money and sales for the sake of it you will not be able to reach the more elusive but more effective brain regions that bring solutions to unexpected problems. Money and sales are part of what has been called extrinsic rewards (Buchanan and Hunczynsky, 2010), which are measured outcomes set and controlled by others. This extrinsic motivation can only work with straightforward, routine-based problems.

Intrinsic motivation frees your brain from the pressure of immediate and short-sighted rewards and allows you to perform better than before. Therefore, this motivation works better for difficult, challenging and complex targets that require our creativity and complex thinking as well. This seems to be the case in an experiment that we have conducted with our students. We separated the students into two groups. Each group had the same target. They needed to enter a room one by one and complete a puzzle, which was the same for both groups. In the first group we asked them to complete the puzzle without promising any reward, saying that we will just count the time that each needs in order to fulfil the target. We let them know that we are grateful to them and will acknowledge their participation in this experiment, which will give us extremely valuable knowledge. In the second group, when we asked them to do the puzzle, we added a sense of competition between them by promising a specific monetary reward to the three fastest students. What do you think happened? According to traditional (rational) management jargon, the pay-for-performance situation (second group) would work better than the non-extrinsic reward situation (first group). However, the results show the complete opposite. The first group of students was 3.3 minutes on average faster rather than the second group. We have repeated this experiment many times, and in the majority of cases we have observed that the promise of a specific extrinsic reward worked as an obstacle in motivating their brains to think in a creative manner. In contrast, when we facilitated the emergence of intrinsic rewards, students performed better with challenging targets. We have also observed that intrinsic rewards are enhanced by a specific overall purpose. Thus, intrinsic motivation needs a purpose, and purpose apparently starts with 'why', deep inside the brain.

Purpose, as a key leadership characteristic, was highlighted by consultant Nikos Mourkogiannis in his 2006 book *Purpose: The starting point of great companies*. Mourkogiannis claimed that in order to become a great leader one cannot afford to be out of sync with those around them about the purpose that motivates them. So, aspiring leaders need to develop a clear sense of their own moral compass to help them guide themselves and others in future career-defining and life-defining decisions. Mourkogiannis suggests there are four main sources of higher purpose in organizations:

- *Discovery.* For people driven by discovery, life is an adventure. The continuous search for the new pushes people to make strong promises and to do their best to deliver on them. Tradition, familiarity and current constraints are the worst enemies of discovery. People and companies guided by this purpose will always go the extra mile to pursue their aims and especially to be the first to serve a new, untapped market need.

- *Excellence.* For those driven by excellence, life is an art form. Their actions do not derive so much from customer demand but from an obsession over the inherent qualities of the product or the service. Their output has to be the best possible, the best available; in one word, perfect. Like artists, their high standards are their guiding light and are non-negotiable. Lowering standards in processes and outcomes is considered a compromise and even a defeat.

- *Altruism.* For those employing altruism as their purpose, life is simply about supporting others. It refers to the personal or institutionalized instinct for providing help to our communities. This caring attitude can be towards customers, employees, suppliers, towns and even the environment as a whole. It is actually about having the well-being of those around you as a top priority and your own selfish needs as a second or third.

- *Heroism.* Those believing in heroism believe that single individuals, or single companies, can and should change the course of history and influence for the better the lives of everybody else. They feel that they possess special skills and they should use them fully to achieve the highest possible results, defying the status quo at any cost. Heroism is a highly competitive attitude, but also one of sacrificing one's personal life for the ultimate reason of being on this planet: to develop that product, to build this company, to close that deal, to solve this problem.

Avoiding having a strong purpose, or adopting one superficially and only because it is fashionable do so, is frequently deeply rooted in an idea running throughout management structures within organizations. This idea, both powerful and outdated, originates in conventional economic thinking and considers people as self-centred and self-interested whose work in organizations is based solely on their various contractual obligations. According to Business Professors Robert E Quinn and Anjan V Thakor (2018), numerous practices within companies misfire, and regularly backfire, because they use this centuries-old logic as their foundational assumption. Purpose, though, goes against this assumption by treating humans more personally and emotionally and taking a long-term approach. As Quinn and Thakor empha-size, as a manager you have a choice:

> You can double-down on that [economics-driven] approach, on the assumption that you just need more or stricter controls to achieve the desired impact. Or you can align the organization with an authentic higher purpose that intersects with your business interest and helps guide your decisions. If you succeed in doing the latter, your people will try new things, move into deep learning, take risks, and make surprising contributions.

A successful purpose manages to interlink emotionally personal and corpo-rate interests. To do so, Quinn and Thakor suggest that, apart from the strategic role of top leadership to envision, discover and communicate an authentic purpose, two more groups within the organization are key: middle managers and positive change agents. Middle management is crucial because of their intermediary role between top management and the employee base (Psychogios *et al*, 2008). Middle managers need to be included, motivated and empowered in order to become active communica-tors and implementers of the company's purpose. Positive change agents (or 'energizers' as Quinn and Thakor call them) are spread within the organiza-tion and people tend to naturally recognize them and trust them. These energizers, when identified and organized in a meaningful network of collaborators, form the fuel of change, inspiring and guiding others towards our purpose.

In our experience, purpose is the starting point of a great leader, a great colleague, a great friend. Again and again we meet managers and directors that are strong in skills and knowledge, they have clear career aspirations, they even have a good track record of results, but lack the internal fire to bring themselves and those around them to a new level. We usually advise those individuals that whatever brought them there will not necessarily make them successful at a higher level in the organization. A move from the tangible to the intangible, from just targeting more sales to enhancing their creativity and complex thinking, is a prerequisite for them to endure the challenges of steering companies and top-level teams. It is needed to keep them going, keep them learning, keep them connecting and keep their brain improving. A higher purpose is not a leadership luxury but a leadership prerequisite.

HOW TO SPOT THE WRONG PURPOSE

We always start our coaching sessions by enquiring about the person's purpose. Through a dialectic process of exploring motives and aspirations we guide our clients to perform a simple diagnostic of their own purpose and of those of people around them. This diagnostic consists of the following four questions in this specific order:

1 *Extrinsic or intrinsic?* If the individual is driven by short-term, overly focused, egoistic and materialistic motives, then they are not on the right path. Extrinsic rewards that follow a narrow-minded 'if–then' logic ('if I do this then I get this') do not allow for the full potential of a true purpose to be realized. A true purpose needs to derive from higher ideals like positive contribution to the lives of the people around us, and from deeper emotions such as love and caring. We help them discover by themselves the potential 'why' of their actions that often is hidden deep inside them.

2 *Complicated or straightforward?* If the individual uses too many words to describe their purpose then something is wrong. Words do not work in our favour here simply because the deeper brain structures that are responsible for commitment, passion and behaviour do not have language capacity. The more words that are used, the stronger the indication that the executive part of the brain in the neocortex has taken control and thus not much motivation is expected. Great purposes are simple to explain and powerful in moving people. Of course, if the individual replies in a fast and straightforward way but mentions extrinsic drivers then we start again. We help them express their purpose in a way that everybody can easily understand.

3 *Forced or natural?* When an individual uses corporate language and/or just repeats the mission, vision and values of the company then we have a good reason to be concerned. A higher purpose is genuine when it comes naturally to the individual and is not imposed in any way. Even when someone is aligned with most of the organizational values this does not mean that a carbon copy of corporate materials will be produced during our enquiry. When this happens, we suspect that our client is just hiding the real purpose behind the official one. After further discussion and some homework people usually expose their drivers and then the real work starts.

4   *Low or high share potential?* Here we explore whether the higher purpose is easily transmittable to other people within and outside the organization or not. If yes, this means that this purpose will not only serve as the leader's guiding light but will engulf internal teams and external customers, suppliers and partners as well. Higher purposes that go viral through a company's people, products and communications have changed not just individuals, but whole industries and societies too. If the purpose is not easily transmittable to others and cannot inspire them then we need to review either the purpose itself or the environment, or both.

ACTION BOX

Individually

Go through the four questions in 'How to spot the wrong purpose'. Keep notes. After you answer each question, consider what you have written based on the discussion in this chapter. Is your purpose in accordance with best practice? If not, what can you do to improve it? Rewrite your answers until you reach a satisfactory result. You can use Mourkogiannis's model to help you define or redefine your purpose.

As a team

You can do this exercise with your team as well, but only if your search for the right purpose is genuine. This will lead to other people being open and sharing their inner thoughts and feelings. Identifying different higher purposes within the team is not a problem. In such cases, discuss with your team how your different purposes can support each other in achieving your organizational objectives. Shared inspiration is what we aim for. We suggest, though, proceeding with the team task after you have identified and defined your purpose on an individual basis.

Purpose motivates you from the inside out to always move forward and upwards, driven by an unlimited passion and a higher principle. But purpose will never help you do anything extraordinary if not accompanied by the brain state of top performance. This state is called 'flow' and on your way to great leadership you need to embrace it fully and enthusiastically.

Flow to greatness

The brain does not perform optimally by default. It can do so under specific circumstances that can be observed and analysed. We can then induce this optimized brain state and achieve what we aim for, accelerate our learning and develop expertise much faster than expected. This realization has revolutionized the way we see performance. It should also make neuroscience the closest 'friend' of all professionals around the globe.

When was the last time you felt and performed at your best? How often and how easily have you found yourself in perfect concentration and in absolute control of the moment? When were you able to absorb and comprehend information fast and deliver effective solutions that made you and your team proud? In short, when and how are you getting in the flow? What most people call 'being in the zone' in neuroscience is called flow. It describes the moments of total absorption that lead to an optimal state of consciousness. Optimal state of consciousness means best possible performance, output and well-being. Mihaly Csikszentmihalyi (2008), the famous Hungarian-American psychologist who first studied this state methodically, devoting his life to it, fittingly stated that:

> [T]he best moments in our lives are not the passive, receptive, relaxing times...
> The best moments usually occur when a person's body or mind is stretched to
> the limit in a voluntary effort to accomplish something difficult and worthwhile.

Almost everyone can recall a moment where, in high adversity, they found the inner strength to do their best, feeling fully empowered, focused and content. In such situations we experience a kind of alternative reality where time is altered, going too fast or too slow, and our sense of our self merges with our surroundings. In short, we are unstoppable! Such a state is critical for leaders to perform amazingly well when needed the most, and create followers that will believe in them whatever the situation.

The future belongs to flow, and many remarkable projects are taking the concept further. The neuroscientist and entrepreneur Chris Berka and her team have created a simple device that you can wear on your head to measure the two key brainwaves involved in flow, alpha and theta (Advanced Brain Monitoring, 2020). According to the *American Heritage Dictionary of the English Language* (2011), alpha waves can be considered as a pattern of smooth, regular electrical oscillations in the human brain that occur when a person is awake and relaxed, and they have a frequency of 8 to 13 hertz.

According to the same source, theta waves have a frequency of 4 to 8 hertz, and they are recorded chiefly in the hippocampus of carnivorous mammals when they are alert or aroused. Alpha waves produce a state of meditative concentration and theta waves extreme relaxation. Imagine measuring your brainwaves while you consciously try to tune them for flow. As soon as you achieve the right measurement in alpha and theta brainwaves through relaxation and concentration, you start dealing with a difficult and crucial situation. You are now in optimized performance. According to Berka, such technologies will allow you to learn, be more creative and master new skills faster than ever. Furthermore, Steven Kotler (2014a) and his Flow Genome Project are committed to uncovering the neurobiological aspects of flow. By looking at the chemical, electrical and structural changes in our brains before, during and after flow, he aims to provide solutions for people to achieve it more often. Kotler (2014b) has found that, apart from changes in brainwaves, flow demands the presence of a complex cocktail of neurochemicals that are collectively responsible for the three main characteristics of rapid problem solving:

- *Muscle reaction times to attention.* This means that our body, as a whole, reacts faster than usual towards a goal.
- *Pattern recognition.* This is the upside of pattern recognition, where our brain is more efficient and effective in reading a situation as accurately as possible.
- *Lateral thinking.* Our brain identifies new solutions by viewing a situation in fresh and alternative ways.

In our experience, the key internal and external conditions for achieving flow are the following.

Internal

- *Higher purpose.* There is no flow without feeling that what you do serves your personal higher purpose. The more you are genuinely, actively and passionately motivated by your purpose, the easier it is to get into flow.
- *Advanced expertise.* Although flow accelerates learning and helps you develop personally even further, it cannot be achieved if you miss the basic skills of your trade. You need to be good to become excellent and perform with excellence. And excellence can be achieved after practice.

Howe *et al* (1998) argued that a talent (like leadership) can become a critical skill mainly because of long and hard work.

- *Low anxiety.* The higher the level of anxiety, the lower the possibility of getting into flow. Bad stress and too much pressure are the worst enemies of getting into the optimal state of flow because they do not allow for the winning combination of meditative concentration, strong emotional rewards and intrinsic motivation to manifest simultaneously. However, no stress at all is equally counterproductive in getting our brains into a flow state. Good stress is a necessary condition for super performance (Conti *et al*, 2006).

External

- *A new challenge.* You will probably not go into flow if you are facing a routine problem. Routine problems put the brain into autopilot and habits are called to finish the job. There is no need to waste additional brain energy if the challenge does not include novel elements that make it exciting and engaging.

- *Some control.* It will also be very difficult to get into flow if you have absolutely no control over the situation at hand. Without empowerment and the ability to take freely decisions to solve the problem, there is little space for finding the inner strength to become the best version of you. A sense of autonomy and empowered contribution go hand-to-hand with the ability to achieve flow. But too much control, especially the cognitively inhibitory and restrictive kind, the one that forms our strictest inner critic, is destructive to the flow state.

- *Increased significance.* The more important the situation, the easier it is to get into flow. Problems with limited implications, regardless of how surprising and novel they are, cannot create the sense of urgency and the feeling of increased impact that can bring about flow naturally and convincingly. We get into flow when it is most needed.

Whatever the next step is in understanding flow better, one thing is certain: our schools, universities, institutions and organizations should move towards becoming flow-friendly organizations where people get into flow individually and collectively to achieve, create and enjoy more. The real question here is: are you becoming a flow leader? That is, are you working on getting yourself into flow and also enabling others to do the same? This is the fastest and surest way to brain-based leadership.

Creativity killed the competition

We have already discussed in the previous chapter the importance of asking questions for keeping a clear mind. However, being curious is not just about maintaining a good view of the world around us. It is also crucial for being creative and bringing new solutions to new problems. In dynamic and unpredictable environments, where rapid changes are the norm, creativity emerges as the number one skill for executives and leaders. It also seems that creativity is the main (if not the only) way to respond to the increased complexity that surrounds us. So, we need to move fast, away from old, obsolete and risky sayings such as 'Curiosity killed the cat'. The only thing that curiosity, and creativity, can really kill is your competition!

An IBM survey of more than 1,500 CEOs from 60 countries and 33 industries around the globe in 2010 found that creativity is considered *the* essential skill for navigating effectively the increasingly complex business environment. CEOs in this study placed creativity as the top required skill, above more traditional business skills such as rigour, management discipline, integrity and vision. This is a remarkable result, since creativity is traditionally considered as a skill reserved for specific professions such as artists, advertising gurus and inventors. Indeed, creativity was rarely a core subject in business studies, seminars and trainings outside marketing communications subjects and innovation courses. A study conducted by Ros Taylor (2013) in the UK across various industries with 100 executive interviews revealed that 70 per cent believed that creativity was less compatible with the workplace and more compatible with arts. Her subsequent survey of 1,000 professionals across the country showed the lowest scores for the statements 'Creativity is at the top of my organization's agenda' and 'Everyone [in my organization] systematically uses a creative tool or technique'. Regardless of how badly CEOs want to have it in their companies, creativity is not part of the DNA of traditional business. However, the worldwide obsession with continuous innovation at all levels has brought about creative thinking, as we have argued above, as the main weapon for leaders to use to deal successfully with complexity. But where do creativity and creative problem solving come from?

The part of the brain that brings the most insightful new ideas is not the part that does most of the analysis and rational thinking. Evangelia Chrysikou (2014), an Assistant Professor of Psychology at the University of Kansas, summarizing research on the neuroscience of idea generation, has argued that lower activity in the executive part of our brain is crucial for creativity. Hypofrontality, which occurs when there is diminishing activity in the prefrontal cortex, leads to fewer constraints, lower attentional focus and the abandonment of rules. The mind becomes more open and new ideas emerge more freely. The quieting down of the executive brain is positively associated with a higher creative output. The less we think about a problem, the more creatively we can solve it. Evidence comes also directly from industry. Mike Byrne, founding partner and global chief creative officer of Anomaly, an unconventional marketing agency with offices in three continents, revealed in *Entrepreneur* magazine (2015):

> I give every idea time to 'bake'…. The best baking always comes when I am doing something else. For some reason it opens up my thinking. I get clarity. I can solve a problem quicker because the baking helps me get to the answer quicker.

It is not uncommon for executives and other professionals that rely on creativity to refer to running, showering, taking a walk and even 'sleeping on it' as great strategies in reaching creative solutions to new challenges. Executive MBA students and a few corporate clients of ours have expressed very similar opinions. This is not to say that not thinking actively about a problem is ignoring it. 'Baking' an idea, as Byrne put it, is not running away from it. On the contrary, quieting our executive brain *after* learning as much as we can about a problem, not before, is crucial for the rest of the brain to offer us the proper relevant solution.

There are two main points in this remarkable brain process. The first is the starting point, called encoding, and it refers to the brain's effort to absorb as much information through the senses in order to solve the problem at hand. The second is the end point, called retrieval, when the brain decides on the best course of action and externalizes this decision through thought entry, meaning the decision enters our conscious mind, and behaviour. In a pioneering study demonstrating this brain process, Bursley and associates (2016) placed participants in an fMRI scan and showed them a series of photos of fictitious animals with an assigned made-up name for each animal. Then they asked participants to remember the animals and their names.

However, some participants were given a distracting task right after the animal/name photos, while others had to respond immediately, without any distraction. Which group remembered more animals and their names correctly? The distracted participants performed much better than the ones who were asked immediately for their response. The study is pioneering because it is the first one to demonstrate 'that a brief, active period of post-encoding offline processing boosts associative learning' and that this can happen when people' minds are consciously and actively pursuing other goals. Furthermore, the study's fMRI results showed that the same brain areas activated in the encoding phase were re-activated in the distraction phase. This means that 'activity patterns present in the left DLPFC during encoding are reinstated during offline processing, supporting the hypothesis that encoded memory representations were reactivated during offline processing'. The brain relives the experience of encoding data, in the form of memories, in order to process information 'as is' and to provide the most intelligent and realistic answer possible. And all these happen outside our awareness spectrum.

In earlier efforts to understand this process, Bursley (2012) stated that offline learning and decision-making perform at their best under two conditions. The more complicated the problem, the better the decision of the subconscious; and also, the higher the motivation, the better the decision of the subconscious. The more the brain cares and the individual is personally invested in a situation, and the more demanding the situation is, the better the result of awake, offline processing.

Amazingly enough, this applies even to sleeping, when the often over controlling executive functions of our brain are mostly absent. It is the same principle at work with 'baking': we go to sleep with our mind full of data and information about a problem that troubles us, and we dream about the solution or we have it ready just before we fully wake up. This seems to be particularly true for brain scientists, since both Adam Gazzaley, the neuro-scientist at the University of California mentioned earlier in this chapter, and John Bargh, the leading scholar on brain automaticity and unconscious processing, experienced dreams that led them to important career break-throughs. Gazzaley vividly dreamed of a video game where the player had to drive a car and at the same time to shoot down signs of a particular shape. Waking up, he realized that could design a video game to help brains improve (Medeiros, 2017). Similarly, John Bargh was mentally occupied with trying to figure out the interplay between conscious and unconscious brain processes when he dreamed of an alligator turning his belly upwards.

He woke up with the epiphany that unconscious processes come first and conscious ones come second in determining decisions and behaviours (Bargh, 2017). Most notably, the psychobiologist Otto Loewi woke up one night in 1920 with an idea that won him the Nobel Prize for Medicine in 1936: nerve cells communicate between each other using chemicals, or neurotransmitters (Yong, 2018).

This phenomenon is not exclusive to brain scientists, of course. We all experience it. Modern neuroscientific research has shed new light on the phenomenon by identifying two specific sleeping phases as crucial for its manifestation (Lewis *et al*, 2018). The first is non-REM, or deep dreamless sleep, where the brain fires millions of neurons in unison in mainly two of its parts, the hippocampus and the neocortex. The hippocampus stores memories of events while the neocortex holds ideas, data and more abstract concepts. During this phase the brain creates categories and extracts essential information. The second is REM sleeping, when the brain severs the connection between the hippocampus and neocortex, and puts both in a flexible state where new connections can form randomly and freely. As Lewis and associates explain:

> The high excitation, plasticity, and connectivity of REM sleep [the second phase] provide an ideal setting for the formation of novel, unexpected, connections within existing cortically coded knowledge. The synergistic interleaving of REM and non-REM sleep may promote complex analogical problem solving.

Both the non-REM and the REM phases are now considered as contributing to the creative outbursts of sleep. The problem is that the world is experiencing what Matthew Walker, Director of the Centre for Human Sleep Science at the University of California and author of the book *Why We Sleep: The new science of sleep and dreams*, calls a 'catastrophic sleep-loss epidemic' (Johnston, 2017). According Professor Walker, 'sleep loss costs the UK economy over £30bn a year in lost revenue, or 2 per cent of GDP'. These are alarming numbers that should drive leaders to take extra care of the quantity and quality of sleep they and their people experience every night. Unfortunately, sleep is often associated with laziness and carelessness. This has to stop. Sleeping can do miracles for mental clarity, decisiveness and creative problem solving alike, especially when it is accompanied by a strong sense of purpose and a deep meaningful engagement with the scope of your work, your team members and your company's brand.

Nevertheless, it seems that the more we squeeze our brains to produce ideas by looking again and again over data, graphs and spreadsheets, the less creative solutions we will be able to find. This counterintuitive proposal goes against the well-established practice of many companies and managers where more power is allocated to the executive part of the brain whenever a difficult and complex problem arises. Paradoxically enough, it is when creativity is needed the most that we minimize our brain capacity to produce it.

The pioneering work of Jennifer Mueller of the University of San Diego on creativity in organizations further supports the above. Her work suggests that creativity is inhibited greatly when executives think in usual business terms. In the popular and highly cited 2012 paper 'The bias against creativity: Why people desire yet reject creative ideas', she and her associates concluded that even if organizations claim to promote creativity and innovation they actually stop it when they require their staff to come up with the best, most appropriate and readily feasible solution. Organizations' inherent aversion to uncertainty and its avoidance will quickly kill any really creative and novel idea simply because managers will be in favour of safer options on the brainstorming board. Thus, there is a bias against creativity: although we declare we want it, we do not really have the mindset and processes in place to accept it. We want it but we actually try to avoid it.

The implications of the above study are immense, since there is an urgent need to change the way we perform idea generation workshops and apply our filtering process, because currently it filters out most of the creative ideas. Too much thinking will inevitably bring into the picture the executive brain and its eagerness to over-analyse, evaluate risks, calculate consequences and predict future outcomes using what is already known to it. It will eventually regress to the familiar, to the risk-free, to the expected. The executive brain is not the right brain to bring and even judge new ideas in their infancy stage. As a result, we need to quiet it down in order to allow all-powerful insights to surface.

In this respect, we suggest the following six-step approach, based on available science and our experience with our clients, to foster innovation and allow creativity to flourish in your organization:

Step 1: Understanding. Here you need to get as familiar with the problem as possible. You need to do the research, read the key data, look at the main graphs, discuss with all key people and consider all possible angles.

In this step you use your analytical brain to get important information that, together with your personal experience, will provide the raw material for your insightful brain to do its job. The team should do the same.

Step 2: Exploration. When you are trying to come up with ideas and discuss them with your team, keep your mind as open as possible. Too much analysis and criticism at this stage does more harm than good.

2a: Individual exploration. Thinking without thinking is important: keep your brain not on the problem but on the background. Shift it into the foreground when needed to discuss it with the team and/or to consider new evidence. 'Baking' the problem, 'sleeping on it', running or exercising to free up valuable brain space, listening to music and shifting your brain to higher thinking can help immensely.

2b: Group exploration. Discuss the problem in a playful and enjoyable way even if it is very serious. Try to keep yourself and the team away from the stress of overplaying responsibility, failure and disaster. Use breaks often, divert team attention when the discussion is over-heated by creating distracting conversations and tasks, and even take a walk together. A smiling, engaging and purposeful team is a creative team almost every time.

Step 3: Decision-making. When the preferred ideas surface, instead of rushing off to implementation consider them carefully. Do they really address the problem? Have we considered all angles? Did we make sure that exhaustion, fear, feasibility, easiness and risk-avoidance did not make us reject the real creative options? Ask for an opinion from trusted sources outside the company because they usually see problems and solutions without the team's internal biases.

Step 4: Implementation. Putting the idea to work nowadays requires as much creativity as its conception. Planning, allocation of resources and coordination of activities are no longer such straightforward tasks as they used to be. Constantly reviewing the process by getting personally involved, always asking the right questions and passionately keeping the purpose alive for these can do miracles in bringing innovative solutions to life.

Step 5: Results. Success or failure is based on the final outcome. Measuring the obvious requires less creativity than measuring what matters. Examine all possible ways to measure the impact of the implemented solutions.

Include as many metrics as possible, both quantitative *and* qualitative. Celebrate successes in order to cement the creative approach to business problem solving. Deal with failures creatively since there are always important lessons learned.

Step 6: Configuration. Embracing creative thinking opportunistically when trying to resolve a specific problem is fine. But the real long-term solution is to develop a culture and structure that constantly promotes creativity. Such an environment is characterized by playfulness, empowerment, constant questioning, appropriate tools and a feeling of increased purpose in everything we do. As the famous experiment by Kempermann and associates (1997) on mice showed, when a group spends time in an engaging, lively and fun environment it creates three times more neurons in specific brain areas than the group that spends the same time in comfortable but empty settings. Transferred to people, such drastic neurogenesis can form the basis for great achievements within companies. Modern innovation and creativity experts, such as Stanford's Tina Seelig (2012), never fail to emphasize the importance of the right environment and culture for creativity, with others, such as Ros Taylor, author of *Creativity at Work* (2013), highlighting the need for specific techniques, tools and procedures for ensuring that a creative process will always be applied.

BOOST YOUR BRAIN

Observe closely the process of idea generation and decision-making next time your team holds a brainstorming session. Did the team use their brains appropriately to promote creativity or did they allow their prefrontal cortex to lower risk and choose familiarity? If this is the case, then come up with (creative) ideas for how to re-approach the whole process to increase the creative output and promote those ideas within the team.

Option 1: In the next brainstorming session choose a provocative idea from the board that works well for solving the problem but that is not popular with the team because of feasibility issues. Ask everyone to pretend for a moment that this is the chosen solution. Observe how the team behaves now and facilitate the discussion accordingly. This will shift their attention from dealing with uncertainty to dealing with an exciting new project without feeling the weight of responsibility. Creativity will be released!

Option 2: In the next brainstorming session ask participants to throw out ideas quickly, trying not to evaluate them at all in this phase. Write all these ideas down and then ask everyone to mark each idea from 1 to 5 based on two variables: actual problem solving and feasibility/comfort. Discuss any discrepancies and reinforce the need for a solutions-oriented view.

Option 3: You can also ask team members to discuss ideas by advocating and supporting *only* their ideas, and avoiding criticizing the ideas of others. Help them to release creativity by focusing on the positive rather than the negative aspects of decisions.

Creativity is neuroplasticity at its best. Identifying new solutions to new (or old) problems, learning new ways, exploring for unexpected insights and challenging the status quo are enabling new neural pathways to be built inside our brains, keeping us effective and competitive for a long time to come.

If memory serves me right

We have observed, and we are sure you have too, that people in organizations with great memory are very much admired. Actually, they are admired everywhere. When someone's memory is readily available to provide a fact, an event, a name or a date from the past in order to help a discussion progress, other people are impressed. We have experienced this ourselves when working with organizations around the world, leading us to believe that improving your memory is a key part in calibrating your brain for better leadership. And we are not the only ones believing it is so. A poll by *Scientific American Mind* in 2014 revealed that the majority of readers participating in the survey prioritized 'cracking cognition' on their wish list of an enhanced brain over building character and even curing disease. Within this category, the most popular ability they wanted to master was memory, topping the whole survey with 40 per cent of total votes. Joshua Foer (2012), author of the book *Moonwalking with Einstein: The art and science of remembering everything*, puts it like this:

The people whose intellects I most admire always seem to have a fitting anecdote or germane fact at the ready. They are able to reach out across the breadth of their learning and pluck from distant patches... memory and intelligence do seem to go hand in hand, like a muscular frame and an athletic disposition.

And it is not just a powerful impression that good memory helps leaders achieve. It is also improved decision-making and thus 'thinking' in our brain-adaptive leadership model. The complex neural process of taking a decision depends partially but crucially on memory. Memory is responsible for bringing back knowledge and past information in order to deliberate effectively on possible courses of action during the decision-making process, according to Antoine Bechara (2011), Professor of Psychology at the University of Southern California. In essence, a missing memory can lead to repeating mistakes and taking wrong decisions that could have been avoided. How many times have you participated in a meeting where the decision was swayed by an important fact or past experience retrieved from memory by a team member? Everyone's appreciation for this team member must have at least doubled in that meeting!

Our brain memory can be classified into two general categories – the short-term or fluid memory, and the long-term or crystallized one (Eichenbaum, 2011). If you recall something that you were doing a few hours ago, or yesterday or earlier, it is associated with your crystallized memory process. A short-term memory lasts for about a minute or so (Burtis, 1982). Short-term memory does not last long, but it is absolutely necessary for processing the actual information as it is received. According to Curtis and D'Esposito, (2003) the DLPFC is responsible for neuro-activity linked to our fluid memory function. The hippocampus, the seahorse-like part of our brains located right under the neocortex and in the limbic system, plays the key role in short-term memory, consolidating memories from the short to the long-term one and in spatial navigation. People with a damaged hippocampus have both memory and orientation problems. Long-term memory provides the appropriate stored data in order to help us think about the information more thoroughly (Burnett, 2016). Long-term memories are stored in the synaptic connections of neurons in our neocortex. In order to be able to remember something long-term we need to fight the natural process of memory degeneration by creating a dynamic and repetitive process between our hippocampus and neocortex. In particular, long-term memory needs to pass through three stages, namely the *encoding stage* (production of memories), *storage stage* (they need to be stored in the hippocampus and then in the cortex) and *retrieval stage* (they need to be recalled when needed) (Klimesch, 2013).

Information that is not revisited tends to get forgotten since the synapses that hold this memory weaken over time. Donald G Mackay (2014), a Professor of Psychology at UCLA, argues that:

> Just as a builder can make a new structure or repair a damaged one, so could the hippocampus craft new memories to replace those that have been degraded... Such rebuilding might occur whenever someone reencounters a forgotten word or a personal anecdote from the past.

This is the way by which more recent and frequent exposure to a piece of information can restore a fragmented memory and minimize memory loss rate. The process for this being done as suggested by Mackay is the following:

1 Information is captured by our senses.

2 It is transmitted to the relevant cerebral part for processing (let's say the visual cortex if it is a visual stimulus).

3 This information travels to Broca's area in the neocortex that stores memories of words.

4 The person cannot retrieve this information due to time degeneration.

5 The person is exposed again to the same information.

6 The new stimulus is transferred to the hippocampus that interacts with the neocortex to recreate the forgotten memory.

7 The reconstituted memory is stored again but with new strength in the neocortex.

Practically, we need to remind ourselves of important information by revisiting original or indirect sources and thus making sure we rebuild and re-strengthen faded memories. Do not leave this to luck and nature!

Another way of powering up our capacity for stronger long-term memory is emotions. By simply caring personally and engaging more in the activities you are involved in, increased emotions will help you retain more information. Feeling detached and bored will have a negative effect on what you remember from a conversation, meeting or activity. This is because by engaging ourselves actively in a situation we care a lot about, we develop autobiographical memories, or memories that are close to our core self. Emotions in such case act as a glue to make sure we remember information, both good and bad, that is important to us. As psychologist Stephan Hamann (2001) suggested, emotional arousal influences memory both during memory encoding, which means attention and elaboration of the information, and also in memory consolidation. In essence, memories supercharged by strong emotions are reactivated more, remembered better and attracted more attention to them. It is imperative then to double-check your motivation, involvement and emotional attachment to your work and to working situations. The more personally you take your job, the more memorable your working moments will become. In a similar manner, the more personally you take your role as leader, the more memorable your leadership attitudes will become, helping you to repeat them or change them. A strong sense of purpose comes in handy here as well.

The third way for improving your memory is by utilizing associations. This technique is used by those competing in memory championships around the world and it is based on the understanding that the stronger and more varied associations with a piece of information in our neocortex are, the easier it will be for us to store it and retrieve it. Memory is a spatial-related cognitive activity. It is crucial for each new piece of information that needs to be stored effectively to be involved in an action that will make it distinctive and irresistible to memory. As eloquently described in Foer's (2012) book about the art of memory, this is because information needs to take as much space as possible within our cortex by creating as many and as strong synapses as possible. A mundane event can become extremely entertaining if we associate it with fun elements that we completely make up and mentally attach to it. For example, if you struggle to remember the job title of a new acquaintance, think of a friend of yours that has the same job title and imagine him riding uphill on a lama or a donkey along with the new acquaintance. This unusual and amusing mental image will increase associations in your brain and will help you remember it easier.

ACTION BOX

The ancient method of the poet Simonides of Ceos, called the 'Memory Palace', was to store memories in a carefully constructed mental image of a familiar structure (house, road, neighbourhood, etc) by attaching to each piece of information to a specific location within this structure.

Try this for yourself. This exercise needs to take place in a silent location. First, choose five new pieces of information you want to memorize. Then close your eyes and mentally walk yourself through your office (or home). See every corridor, corner, desk and meeting room in your floor in all possible details, but empty of people and noise. After you finish and you are satisfied with the clarity of your memory palace, start mentally walking around it again, placing the five items (images, names, notions, words, whatever they are) in a specific order in different locations within the office. You have to vividly put them down in each distinctive place. To increase memorability you need to also create a unique activity for each item in each location that involves a verb and a person or an item actually doing something. Remember, the more unorthodox, entertaining and edgy the act, the more successful you will be in storing the memory.

After you finish, take a short break and then relax, close your eyes and try to revisit your memory palace, mentally walking around it and finding the items you wanted the remember stored in the places you left them. Practice makes perfect, so the more you play this game the better you'll become. Just make sure you mentally clean your palace before storing new items.

Memory has to serve a leader well. Improved memory will lead to better decisions, more meaningful conversations and overall increased performance. Most importantly, it will dazzle the people around you and help you to enhance your charisma and influence. Furthermore, memory can increase your brain's ability to adapt, which is equally critical for leaders nowadays.

Interestingly enough, though, forgetting is a crucial element for learning and subsequently remembering what really matters. In his book *How We Learn: The surprising truth about when, where, and why it Happens*, Benedict Carey (2015) highlights that forgetting, probably counter-intuitively, enables and deepens learning by filtering out distracting information. Furthermore, forgetting allows for memory breakdown that, after reuse, strengthens the retrieval and storage of information. Bjork and Bjork (2019), the leading experts on the role of forgetting in learning, have summarized three main methods for enhancing learning through forgetting. These methods are:

1 *Changing environments*. Change the sensory context from when some yet-to-be-learned material is studied to when this material is re-studied. Companies should organize the same training sessions in alternative locations in order for the learned material to stick better.

2 *Increasing delay.* Separate, or space out, when something is studied from when it is tested or restudied. Companies should avoid teaching all possible materials on a subject at the same time. They should spread the topic over time for better results.

3 *Interleaving.* Instead of blocking, which is learning in homogeneous subject blocks, we should study or practise mixed elements of to-be-learned knowledge with other material. While learning a subject, people in companies should be exposed to the desired subject together with other subjects at the same time.

These methods have been reported frequently in relevant research as forming a robust framework that helps leaders go deeper into the science of learning, remembering and forgetting. When following these methods, people tend to feel that they do more forgetting than learning. But in the long term, the studied material leads to better memory retention/retrieval and to increased application of the learned skill. Unfortunately, when companies are pressed for time and costs, which is only too often, they do not follow the latest insights on learning, committing the same mistakes again and again. Brain science leads the way and the leaders that utilize it first have a clear competitive edge.

There is one last point concerning memory that needs to be addressed. And it is an important one. It is the fact that memory is far from accurate and highly malleable As Dr Julia Shaw (2016), author of the book *The Memory Illusion: Remembering, forgetting, and the science of false memory*, notes, a memory is a network of brain cells that stretches across brain regions and is constantly updated. Although this is an important neural process, allowing us to learn and to problem-solve, it can be highly manipulated. Essentially, every time we tell a story, we change the memory associated to it by inserting new details, adding elements we maybe picked up elsewhere, or creating new conceptual connections which could be both inaccurate and misleading (Lunau, 2019).

This de-romanticized view of memory strips away much of its credibility as a process and as an outcome. Memories are not always as real as our minds want us to believe. They are the product of dynamic and sensitive neural firings, which do not simply store and retrieve data in a disk like computers do, but which recreate and relive an experience based on emotions, impressions, new information and hidden brain agendas. Indeed, when leaders intend to use a past event as a key point in their storytelling or when constructing a crucial argument, they need to consider the facts very carefully and use multiple and reliable sources to double-check all vital information. However, there is another important lesson here. When a business situation proves to be damaging for the team, creating a negative emotional reaction and a destructive memory within the company, brain-adaptive leaders can turn it into a constructive learning opportunity by re-framing the story later on. The nature of memory is an actual strength for humanity since it allows for revisiting a situation with a different perspective and changing, not facts, but our attitude towards that situation. By recreating the story for our team, we recreate memory too. This approach, which is used in psychological trauma therapy with positive results, is a strategic tool for leaders in their efforts to convey courage and resilience to teams with low morale.

There is a last but important lesson in relation to our memory system. This is related to the context within which the information and therefore memory was created and stored. In a famous experimental study (Godden and Baddeley, 1975), two groups of people were asked to learn some information. The first group was asked to learn the information by simply listening to it in a room, while the second group was asked to do the same underwater in scuba-diving suits! Afterwards, the researchers tested both groups either under the same situation (regular room or underwater) or in different ones (those that learned the information underwater were tested in a room, and those that learned the information in the room were tested underwater). The results showed that these that received the information and were tested in the similar situation performed much better than those tested in different situations. This study shows that the type of the context does not matter (at least that much) to the process of learning and long-term memory development, but it is important it remains the same in case of memory retrieval. In other words, putting someone in the same context helps him to activate long-term memory and facilitate the learning process (Burnett, 2016). This finding has enormous implications in leadership development. As we have argued above, putting yourself in a leadership context, in a situation where you need to lead others to achieve something collectively, provides an excellent opportunity to retrieve leadership behaviours and facilitate the development of your leadership capabilities. In addition, a memory of a leadership context seems to influence the perception of followers about their leaders. In a study, Boyatzis and his colleagues (2012) used fMRIs scans and asked subjects to recall past events with two types of leaders, namely resonant and dissonant ones. Their findings show that memories related to resonant or dissonant leaders activated specific parts of the neural system of the brain, affecting in turn positively or negatively additional cognitive and emotional aspects related to the one or the other part of the brain. These results support further the idea that the more we expose ourselves or others in similar leadership experiences, the stronger the memories that can be created, stored and retrieved, influencing both leaders' and followers' behaviours.

Adapt, bet and grow

The Economist publishes annually its most important predictions for the year to come. The esteemed 'The world in [year]' publication includes a number of predictions in economics, business, politics and technology for the following year. In its 25th year special edition *The World in 2011* (*The Economist*, 2010), the prediction for the country Libya in the Middle East and Africa section stated:

> Muammar Qaddafi has held power for 40 years and will certainly complete 41. By suppressing opponents and undermining rivals, he has removed all significant threats to his rule; the only credible successor is his own son, Saif al-Islam.

Knowing with hindsight that 2011 was the year that General Qaddafi and his family were ousted from power and Qaddafi himself was killed by the rebels, one can only be startled by the confidence in that particular prediction. We've seen in the previous chapter that such an attitude fits the profile of a hedgehog better than a fox. It also portrays a simple but powerful assertion for the world we live in and the corresponding mindset leaders need to adopt to survive. The level of complexity we experience today, based on the high number of interrelated forces shaping simultaneously all aspects of our lives, supports a very different approach in decision-making for leaders of all types. From the outdated approach of choosing one single and rigid possible future and behaving accordingly we need to move to a more flexible, open and careful approach of choosing multiple ways forward and pursuing them with not necessarily equal, but definitely critical attention. This new approach is consistent with the popular writings of modern philosopher Nassim Nicholas Taleb (2007, 2012) on randomness, black swans and antifragility. In particular, Taleb claims that randomness and unpredicted events are far from being under our control and ability to foresee. Stability, routine, incremental and planned improvement is something that humans prefer and target rather than radical and sudden change. However, this situation, given that unpredicted events (black swans) will come sooner or later, makes us more fragile and sensitive to their impact. In contrast, our constant exposure to small variations (avoiding from time to time routinized processes of actions) can make us potentially more resilient or, as Taleb argues, *antifragile*. This approach calls for what has come to be known as small or purposeful betting. This is the best cognitive strategy we know of and apply for steering people and organizations through the dangerous waters of unpredictability.

With our digital selves, mobile devices and social media dominating the collective global conversation, with political uncertainties and conflicts constantly bursting around the world and with a start-up culture challenging and dethroning industrial behemoths on a daily basis, leaders have little room to be certain about the future. This creates a serious issue with strategy, personal and organizational, since strategy, a critical aspect of leaders' thinking, traditionally relied on predictions of, and decisions about, the future. If the future is impossible to foresee, how can we make any solid decisions on investments, resource allocation, preparation and activity planning? By employing the mindset of the purposive bets, we can deal with our chaotic reality more effectively and at the same time lower the overall risk of failure.

Purposeful bets are defined by author and consultant Frans Johansson (2012) as the actions we take regardless of our ignorance of their effectiveness, and by entrepreneur and author Peter Sims (2011) as actions of low risk, intended to unearth and try ideas out. Since high uncertainty defines all areas of human endeavour and we cannot stay idle, it is advisable to take as many diverse actions as possible at the same time, investing in – or betting on – their success. This will eliminate the possibility of a fatal blow if one or some of them do not work out. The leadership mindset deriving from this approach is one of braveness, hyperactivity, connectedness and optimism. The executive part of our brain does its usual analysis, forecasting and decision-making by quieting the amygdala's fear of not surviving. If many small bets fail, all is not lost. However, if just one succeeds it could make a huge difference in the overall outcome.

The five key steps in placing purposive bets according to Johansson (2012), which we have also been actively applying, are:

1 *Place many bets.* Playing only in one arena and envisioning just a single possible future can be catastrophic. Engaging in multiple activities simultaneously and always discovering and following new ways to pursue goals are necessary for success. For example, when faced with a specific challenge, do not search for the one and only 'golden solution', but engage in exploring various alternatives. We recommend choosing a number of actions that can co-exist in implementation rather than being mutually exclusive to each other.

2 *Minimize the size of the bets.* The main characteristic of bets is that they are low-risk. This means that although we do commit resources and time to implementing them we base the financial and strategic survival of our organization on them not individually but collectively. We have observed that committing all resources in one solution can be fatal both for a company and for managers' careers. We need to move away from the all-powerful magic idea that can solve a problem instantly (but expensively too) to a manageable portfolio of more cost-effective solutions.

3 *Take the smallest executable step.* Since bets are used to examine ideas in practice and many of them simultaneously, they should be approached with caution. A methodical approach should be adopted where every task leads to the next only when the validity of the previous step is established. Allocating different steps to different people, using external resources (researchers, consultants) to help verify the most important of them, and using your closest team to evaluate progress and approve next steps are key actions that we suggest to our clients.

4 *Calculate affordable loss.* Instead of analysing the possible return on investment (ROI) of every bet in order to decide whether to proceed with it or not, give the go-ahead based on how much is acceptable to be lost in each one. ROI can stop you pursuing activities even with very low investment requirements without really ensuring that the ones chosen will actually deliver as expected. So, it is better to stop having expectations that are anyhow highly uncertain and focus instead on getting bets off the ground by investing as little as possible in each of them.

5 *Use passion as fuel.* Placing numerous bets constantly, and following them closely, even witnessing many of them dying out, requires a very strong commitment. As already discussed in this chapter, a higher purpose, passion and intrinsic motivation can keep you going, learning from the failed bets fast and using every possible success as the fuel for your continuous efforts. In our experience, highly driven teams are the ones

that do the most and overcome any mishap on the way. Actually, we have worked with leaders that were happy to see bets going wrong because they felt they were closer to the ones that would work well.

The mentality of foxes, the inner power to harness randomness, the ability to manage multiple purposive bets at the same time while embracing creativity, flow and a higher purpose are the core elements of the mindset that contemporary leaders need to employ. This mindset, which promotes positive neuroplasticity and neurogenesis, is remarkably similar to the growth mindset suggested by Dr Carol S Dweck. A mindset greatly affects our behaviour, relationship with success and failure and ultimately our aptitude for happiness and well-being. Dweck (2012) has convincingly explained that there are two main mindsets that people adopt: the fixed one and the growth one. The fixed mindset believes in set abilities, intelligence and personality and does not believe in meaningful change. It results in a life overly dependent on external circumstances rather than on your own actions; in psychology this is called an external locus of control. On the other hand, a growth mindset promotes the freedom of choosing who you are by developing and improving your abilities. Intelligence, personality and capabilities are not static. Failure is just another opportunity to learn and grow, while success depends on your own efforts. You can ultimately change and become a better you. This is called an internal locus of control.

The growth mindset is using neuroplasticity in the best possible way, while the fixed mindset is using it in the worst possible way. We have seen again and again that leadership thrives on a growth mindset while it degrades in a fixed one. Which is your mindset?

Keep in mind

Our brain does not remain biologically unchanged during our professional life. It can do more or it can do less, depending on how we use it. Leaders need to take advantage of the brain's neuroplasticity and learn faster, perform better and achieve more. Memory is a critical part of brain function and can dramatically help in developing our leadership capabilities. Adopting a higher purpose, achieving flow, using creativity in all situations, improving memory and using a diversification, adaptive and growth mindset can help leaders remain always sharp and on top. Your optimized performance will create more devoted followers for you.

References

Advanced Brain Monitoring (2020) Official website, www.advancedbrainmonitoring. com (archived at https://perma.cc/7WD8-GGT2)

American Heritage Dictionary of the English Language (2011) 5th edn, Houghton Mifflin Harcourt Publishing Company, New York

Bargh, J (2017) *Before You Know It: The unconscious reasons we do what we do*, William Heinemann, London

Bechara, A (2011) The somatic marker hypothesis and its neural basis: Using past experiences to forecast the future in decision making, in *Predictions in the Brain: Using our past to generate a future*, ed M Bar, pp 122–33, Oxford University Press, Oxford

Bjork, RA, and Bjork, EL (2019) Forgetting as the friend of learning: Implications for teaching and self-regulated learning, *Advances in Psychology Education*, **43**, pp 164–67

Boyatzis, RE, Passarelli, AM, Koenig, K, Lowe, M, Mathew, B, Stoller, JK and Phillips, M (2012) Examination of the neural substrates activated in memories of experiences with resonant and dissonant leaders, *The Leadership Quarterly*, **23**(2), pp 259–72

Buchanan, DA and Hunczynsky, AA (2010) *Organizational Behaviour*, 7th edn, Prentice Hall, Upper Saddle River, NJ

Burnett, D (2016) *Idiot Brain: What your head is really up to*, WW Norton and Company, New York

Bursley, JK (2012) Unconscious learning: James Bursley at TEDxEnola, www. youtube.com/watch?v=ghPX9NhPqpg (archived at https://perma.cc/C397-PP9D)

Bursley, JK, Nestor, A, Tarr, MJ and Creswell, JD (2016) Awake: Offline processing during associative learning, *PLOS ONE*, **11**(4): e0127522

Burtis, P J (1982) Capacity increase and chunking in the development of short-term memory, *Journal of Experimental Child Psychology*, **34**(3), pp 387–413

Carey, B (2015) *How We Learn: The surprising truth about when, where, and why it happens*, Random House Trade Paperbacks, London

Chopra, D and Tanzi, RE (2013) *Super Brain: Unleash the explosive power of your mind*, Rider Books, London

Chrysikou, EG (2014) Your fertile brain at work, *Scientific American Mind*, Special Collector's Edition on Creativity, **23**(1, Winter), pp 86–93

Conti, R, Angelis, J, Cooper, C, Faragher, B and Gill, C (2006) The effects of lean production on worker job stress, *International Journal of Operations and Production Management*, **26**(9), pp 1013–38

Csikszentmihalyi, M (2008) *Flow: The psychology of optimal experience*, Harper Perennial Modern Classics, London

Curtis, CE and D'Esposito, M (2003) Persistent activity in the prefrontal cortex during working memory, *Trends in Cognitive Sciences*, 7(9), pp 415–23

Dweck, CS (2012) *Mindset: How you can fulfill your potential*, Robinson, London

Eichenbaum, H (2011) *The Cognitive Neuroscience of Memory: An introduction*, Oxford University Press, Oxford

Entrepreneur (2015) Creative genius: Habits and tips from inventive people in the business, April, pp 56–60

Foer, J (2012) *Moonwalking with Einstein: The art and science of remembering everything*, Penguin Books, London

Fortune (2014) CEO 101: Business leaders share their secrets to success, *Fortune*, 17 November, pp 46–53

Fox, E (2013) *Rainy Brain, Sunny Brain: How to retrain your brain to overcome pessimism and achieve a more positive outlook*, Arrow Books, London

Godden, DR and Baddeley, AD (1975) Context-dependent memory in two natural environments: On land and underwater, *British Journal of Psychology*, 66(3), pp 325–31

Hamann, S (2001) Cognitive and neural mechanisms of emotional memory, *Trends in Cognitive Sciences*, 5(9), pp 394–400

Hood, B (2014) *The Domesticated Brain*, Pelican Books, London

Howe, MJA, Davidson, JW and Sloboda, JA (1998) Innate talents: Reality or myth?, *Behavioral and Brain Sciences*, 21(3), pp 399–407

IBM (2010) IBM 2010 global CEO study: Creativity selected as most crucial factor for future success, IBM Press News, www-03.ibm.com/press/us/en/pressrelease/31670.wss (archived at https://perma.cc/8MD6-ECYF)

Johansson, F (2012) *The Click Moment: Seizing opportunity in an unpredictable world*, Penguin, London

Johnston, I (2017) 'Catastrophic' lack of sleep in modern society is killing us, warns leading sleep scientist, www.independent.co.uk/news/sleep-deprivation-epidemic-health-effects-tired-heart-disease-stroke-dementia-cancer-a7964156.html (archived at https://perma.cc/5L8Q-EUYS)

Kempermann, G, Kuhn, HG and Gage, FH (1997) More hippocampal neurons in adult mice living in an enriched environment, *Nature*, 386(6624), pp 493–95

Klimesch, W (2013) *The Structure of Long-Term Memory: A connectivity model of semantic processing*, Psychology Press, Hove

Kotler, S (2014a) *The Rise of Superman: Decoding the science of ultimate human performance*, Houghton Mifflin Harcourt, New York:

Kotler, S (2014b) The science of peak human performance, https://time.com/56809/the-science-of-peak-human-performance/ (archived at https://perma.cc/GUN8-Q7SM)

Lewis, PA, Knoblich, G and Poe, G (2018) How memory replay in sleep boosts creative problem-solving, *Trends in Cognitive Sciences*, 22(6), pp 491–503

Lunau, K (2019) A 'memory hacker' explains how to plant false memories in people's minds, www.vice.com/en_in/article/8q8d7x/memory-hacker-implant-false-memories-in-peoples-minds-julia-shaw-memory-illusion (archived at https://perma.cc/8XCK-SJVB)

Mackay, DG (2014) The engine of memory, *Scientific American Mind*, 25(3, May/June), pp 30–38

Medeiros, J (2017) How to 'game your brain': The benefits of neuroplasticity, www.wired.co.uk/article/game-your-brain (archived at https://perma.cc/UHT3-7VKX)

Mourkogiannis, N (2006) *Purpose: The starting point of great companies*, Palgrave/Macmillan, Basingstoke

Mueller, JS, Melwani, S and Goncalo, J (2012) The bias against creativity: Why people desire yet reject creative ideas, *Psychological Science*, 21(1), pp 13–17

Pink, D (2009a) *Drive: The surprising truth about what motivates us*, Riverhead Books, New York

Pink, D (2009b) The puzzle of motivation, TED Talks, www.youtube.com/watch?v=rrkrvAUbU9Y&t=4s- (archived at https://perma.cc/6QYG-JAWP)

Psychogios, GA, Alexandris, K, and Onofrei, A (2008) Addressing individual and organizational factors influencing middle managers' synthesising role in knowledge creation and diffusion, *International Journal of Learning and Intellectual Capital*, 5(2), pp 208–22

Quinn, RE and Thakor, AV (2018) Creating a purpose-driven organization, *Harvard Business Review*, July–August, pp 78–85

Scientific American Mind (2014) Upgrading the brain, *Scientific American Mind*, Special issue: The future you, 25(6, November/December), pp 8–9

Seelig, T (2012) *inGenius: A crash course on creativity*, Hay House, Inc, Carlsbad, CA

Shaw, J (2016) *The Memory Illusion: Remembering, forgetting, and the science of false memory*, Random House, London

Sims, P (2011) *Little Bets: How breakthrough ideas emerge from small discoveries*, Random House, London

Sinek, S (2009) Start with why: How great leaders inspire action, TED Talks, www.youtube.com/watch?v=u4ZoJKF_VuA&t=5s (archived at https://perma.cc/4UJK-GHWY)

Sinek, S (2010) *Start With Why: How great leaders inspire everyone to take action*, Penguin, London

Sinek, S (2014) *Leaders Eat Last: Why some teams pull together and others don't*, Penguin, London

Taleb, NN (2007) *Black Swan: The impact of the highly improbable*, Penguin, London

Taleb, NN (2012) *Antifragile*, Allen Lane, London

Taylor, R (2013) *Creativity at Work: Supercharge your brain and make your ideas stick*, Kogan Page, London

The Economist (2010) The world in figures: Countries, *The World in 2011: 25-year special edition*, *The Economist*, London

Yong, E (2018) A new theory linking sleep and creativity, www.theatlantic.com/science/archive/2018/05/sleep-creativity-theory/560399/ (archived at https://perma.cc/73Z5-6C27)

SUMMARY OF PILLAR 1: THINKING

Leadership lessons

Save power for your brain

Higher-level thinking, strong values and immediate feedback are three key aspects for strengthening the leadership willpower muscle.

Save power from your brain by:

- dealing with ego depletion;
- dealing with burnout;
- dealing with multitasking;
- developing resilience.

Clear your mind

Being aware of biases that can blind you when making decisions.

Use the PAUSE framework to help you overcome biases:

P – Pay attention to what's happening beneath the judgement (event vs interpretation).

A – Acknowledge or identify your reaction/interpretation/judgements.

U – Understand other possible reactions/interpretations/judgements.

S – Search for the most constructive/empowering/productive way to deal with the situation.

E – Execute your action plan (act consistently with what makes the most sense).

Actively asking questions brings more energy to the rational part of the brain.

Four behaviours when asking questions:

- the star enquirer;
- the good listener;
- the evil interrogator;
- the grumpy loner.

The brain hijacking mechanism threatens clear thinking, with important consequences like erratic behaviour, impaired decision-making and performance, etc. Hijacking can be managed through a system of distractions and rewards.

Focus on performance

Neuroplasticity can help us develop new skills and capabilities. Leadership can be developed by actually practising leadership. The more you are involved in leadership tasks, the more chances you have to enhance this cognitive capability.

Strong purpose can help leaders to see the direction and understand better the decisions made and the decisions that will be made.

Creativity and performance can be enhanced by:

- understanding the context of the problem;
- avoiding too much analysis;
- using the simplest analytical tool;
- constantly reviewing the implementation process;
- continuously evaluating the results;
- repeating the process as many times as possible.

Memory plays an important role in leadership development. The more you put yourself in a leadership situation, the more you enhance your crystallized leadership capabilities stored in the long-term memory system of your brain. And the more often you retrieve these memories, the more chances you have to become better and better.

Deal with complexity by employing purposive bets:

1 Place many bets.

2 Minimize the size of the bets.

3 Take the smallest executable step.

4 Calculate affordable loss.

5 Use passion as fuel.

Emotions

04

More emotion, better decisions

ALWAYS STAY COOL

He considers himself a great manager. He has held directorial positions in many companies and his track record of results has been stellar, apart from the occasional mishaps that he quickly resolved. He has always been very careful in considering options before making a decision, as well as involving the right people. His professional standing and general reputation among his peers is outstanding. He has achieved all this by following a simple rule he learned when he was young: always stay cool, no matter how critical the situation is.

His motto is that business is only numbers and that it does not include emotions. His strong belief is that emotions, both good and bad, can influence the decision-making process and alter what should be a clear and logical outcome, the kind of outcome that markets and shareholders prefer. He has learned to reserve his emotions for the very few moments in life without any risk: a family celebration, a concert by his favourite band, watching sports alone at home. His office is an emotions-free zone and he is very proud of it. He considers his thinking as mathematically oriented, rational and objective and this has served him well in his career. He uses this principle not only on himself, but for choosing, evaluating and mentoring members of his team as well. Until now, that is.

Two months ago he got a very important call. He is one of the candidates for the top position in the company: the CEO. This is huge for him but, as always, he treated the news with outer calmness. The process of selecting the new CEO will be long and challenging, and he needs to maintain his composure. He is confident that everything that brought him here will make him succeed at the next level as well.

The company has hired an external consultancy agency to run the process of selection. As expected, this includes interviews, tests, role-play and meetings with different people within and outside the company. For most of the process he felt that things were going fine. Then the final challenge comes. It consists of a task asking candidates to develop a restructuring plan for a large part of the company. How would they approach it? What would be the main areas of concern and which would be their priorities? It is obvious that this is what the shareholders were considering doing with the company and they wanted to make sure that the new CEO would tackle it successfully. He started dealing with the task by applying all his knowledge, experience and known cool attitude to every aspect of the project. He identified the market changes that led to the need for restructuring, he evaluated what needed to be done with the current structure, he set new objectives and structures, he estimated budgets, and he developed an effective implementation plan. He even considered negotiations with external partners and replacements or transfers of current key personnel. It all looked clean and clear to him.

When the announcement came that he was not the selected candidate he was shocked. Even more so as he was told that he underperformed in the last task. How could this be? He did everything by the book and it all made business sense. He still cannot fully understand the feedback, when especially it stated that the way that organizations and individuals should understand the role of CEO is not as chief executive officer, but as 'chief emotional officer'. It continued that a top manager, when faced with such monumental challenges as a restructuring programme, should consider first his/her and other people's emotions, attitudes and persuasions in all analyses and plans. Business, it seems, is not only about numbers. It is firstly about people, because people work with numbers and make decisions (right or wrong) in businesses. Secondly, people do have emotions. A manager should be able to recognize and harness emotions in order to respond accordingly to demanding situations. After this incident, he always tried to consider emotions as part of his core job, both his and those of others. He understood well that there is no situation (professional or personal) not dominated and directed by emotions.

There has been a major misinterpretation of human nature. The advent of science and Renaissance-inspired thinking after the Middle Ages proclaimed cognitive processes as the pinnacle of human experience, relegating emotions to animalistic states. Bashing emotions out of mental processes ensured objective thinking. 'Stay cool', 'don't be emotional', 'you are overreacting' and other phrases like these showcase the aggressive and hostile attitude towards emotions from our modern societies, and they have dominated the traditional jargon of management science since its emergence and theoretical framing by names like Taylor (1911), Fayol (1930, 1949) and Weber (1947). The main message was: *the less the emotion, the better the decision.* But can humans live, think and act without emotions? Are emotions evolutionary leftovers that diminish our rational thinking? Should we try to suppress them to become better leaders? Neuroscience answers all these questions with an emphatic '*No!*' and the clinical case of a patient called Elliot can help us explain why.

The emotion-run brain

Elliot, in his early thirties was diagnosed with a fast-growing tumour in the midline area of his brain, which was pushing both frontal lobes upwards. After successful surgery to remove it, Elliot emerged healthy but changed. Although he was able to think logically and calculate decisions rationally, his actual behaviour was antisocial and far from normal. This change, and similar ones observed in other such cases, led to monumental insights into the role of emotions in decision-making and overall in a healthy human life. In a nutshell, it revealed that emotions are at least as important as any other brain function in the way we take decisions and act on them. Decisions without emotions are not just wrong, but they may also be dangerous.

Elliot's case, together with that of the famous Phineas Gage from the 19th century, whose skull was penetrated by a hot rod while working on a new railroad in Vermont, are described analytically in Antonio Damasio's seminal 1994 book, *Descartes' Error*. Damasio, a renowned neuroscientist/neurobiologist, made the following startling statement in his book: 'Reduction in emotion may constitute an equally important source of irrational behavior'. He convincingly claimed that when emotion is absent there is a counterintuitive connection with distorted behaviour, which reveals the sophisticated but strong dependence of vigorous reasoning on healthy emotions.

The remarkable insight in Elliot's case, and similar ones from medical history, is that while his high-level cognitive functions were intact after the surgery, his decision-making and behaviour was leading to long-term isolation from the people around him. When subjected to tests on perceptual ability, past memory, short-term memory, new learning, language, arithmetic and motor skills, Elliot performed exceptionally well. He appeared to be totally normal and ready to perform his daily personal and professional tasks like everybody else. But his actual behaviour was catastrophic for both. How can somebody pass all those tests, appear totally logical but then perform so badly socially and personally? And why is this a phenomenon that neuroscience confirms again and again (Miller *et al*, 2008)? The answer is as simple and as direct as can be. The ability to use emotions in decision-making and behaviour is as important as, and sometimes even more important than, the ability to use logic. Evolutionarily speaking, emotions developed much earlier in our long history and played a crucial role in making choices. The executive brain, as the latest addition in our skulls, depends greatly on emotions to drive and direct its decisions. Elliot, Phineas Gage and other cases in medical history had actually damaged their neurological paths that inject emotions into their choices and this led to long-term disastrous behaviours. Switching off your emotional brain when taking managerial and business decisions is not only neurologically impossible it is also a very dangerous thing to attempt.

The biggest impact of emotions in decision-making is morality. This is the ability to weight a decision in its later stages based on the possible impact on ourselves and others. It is close to empathy, since it helps us share the feeling that an action can cause. People with damaged emotional neural pathways (amygdala–limbic system and hypothalamus–sympathetic response) cannot use their moral compass to guide their decision-making process and behaviour, and thus end up making wrong choices both for themselves and others. Actually, there are two medical terms for people that do not have the ability to use empathy and thus take only cold-blooded decisions that eventually work against them and society: psychopaths and sociopaths. Jon Ronson (2011) in his well-known book *The Psychopath Test: A journey through the madness industry* and before him Paul Babiak and Robert Hare (2006) in their famous book *Snakes in Suits: When psychopaths go to work* argue that reliable characteristics of a psychopath are an absence of empathy, remorse and loving kindness. Amazingly enough, those are the elements that traditionally schools, MBA programmes and companies tried to push out of future professionals and executives while boosting their cold-blooded, calculative reasoning. Are we trying to make psychopaths out of managers, leaders and people in general? If we are then we seem to be very successful, as a study has shown than 20 per cent of CEOs can be diagnosed as psychopaths, the same portion as in prison populations (Australian Psychological Society, 2016). This is a very worrying fact since the presence of psychopaths in normal populations is just 1 per cent. To spot a psychopath, these are the main characteristics to look for, according to Tomas Chamorro-Premuzic (2019), Professor of Business Psychology at University College London and at Columbia University: constant need for validation and recognition by others, heightened self-centeredness, high levels of entitlement, inability to feel regret and guilt, and constant revenge-fulness. Modern leaders need to move away from such archaic mindsets and be aware of neuroscience-based realities about what make us great in companies and in life.

Morality is a complex emotion. It can be about compassion, but also about fairness and respect. Furthermore, it can be about keeping promises and, especially, about not lying. Moral emotions in humans are multifaceted and dynamic, and have evolved under two premises: 'we' is more advanta-geous than 'I', and I am not more special than my peers (Tomasello, 2019). Both are key components of collective success in organizations. At the other extreme, true evilness can be viewed more as a dysfunction of the brain rather than a state of healthy neurons firing normally on a daily basis. Psychopathy, sociopathy, specific brain tumours and damaged prefrontal cortices cause people to perform malevolent actions reliably and repeatedly. A healthy brain is a moral brain.

Apart from enhancing our decision-making, emotions are significant for leadership because they are the bases of motivation. The Latin root of the English word emotion is 'to move' and this is in plain sight since the word itself includes the term 'motion'. We probably rarely think of it like that but actually the word 'emotion' implies that emotions are the foundation of all our movements. Without emotions we would stay idle, not having the drive to do something, anything. This is why scientists, consultants and entrepre-neurs, like Chip Conley (2012), equate emotions with energy and motion or, as he says, 'emotions are vehicles for transforming or moving your life'. Positive or negative emotions spring us into action to either go towards or retrieve ourselves from a situation. Elaine Fox (2013) has elaborated on these two basic brain systems, arguing that they explain much of our human behaviour. The amygdala, as the fear/emergency/avoidance/pessimism centre and the nucleus accumbens as the pleasure/excitement/acceptance/optimism centre drive most of our movements away from or towards situations. Although these deeper brain structures might be responsible for the emer-gence in humans of what have become known accordingly as the approach and the avoidance motivational systems, their neural footprint is wider than those two brain regions. In a review of available research on the subject,

Kelley and associates (2017) confirmed the expression of these two motivational systems as an asymmetrical activation of the left and the right prefrontal cortices of the brain. When the left side of our prefrontal cortex fires more than the right then the brain intends to approach a situation, while it intends to avoid a situation when the opposite happens. This 'lateralization' ability of our brains has a very important reason to exist in evolutionary terms (Kelley *et al*, 2019):

> These laterality effects are presumed to have evolved to increase an organism's neural capacity and processing efficiency. These laterality patterns may prevent the simultaneous initiation of two antagonistic responses through inhibitory connections between the hemispheres. That is, cerebral lateralization would prevent an organism from simultaneously initiating an approach and avoidance response.

Leaders and managers need to be familiarized with how those systems affect their and other people's behaviour if they are to achieve higher level decisions. This is particularly important because brain intentionality, meaning the subconscious and very fast decision of the brain to give its support or not to a situation, will form the basis of a person's thoughts and actions towards a situation. Approach brain motivation/intentionality is associated with positive emotions while avoidance motivation/intentionality is associated with negative emotions, with anger being the only emotion that has been found to belong to both systems (Kaczmarek *et al*, 2019).

Emotions run our brain. They are usually classified in three categories. First, the fleeting emotions we feel in any given moment Second, the traits, which are emotions that have a long-term presence. Third, the moods, which are somewhere between the first two. Neurologically speaking, the concept of emotional styles is closer to understanding how our brain wiring affects our decisions and behaviours through emotions, and so it is the best place to start our emotional journey.

Emotional style

Psychology has long dealt with emotions but, as with everything else, it did not have the ability to look into the brain to see what exactly was happening and thus could not provide us with as accurate observations as we have today. The advent of neuroscience and the accompanying technology allowed us to peer into the brain to observe how neurons fire and connect with each other. Research has revealed the neurological basis of emotional predispositions and Richard Davidson (Davidson and Begley, 2012), famous for his research on Tibetan monks' brains, has developed a specific categorization of what he calls emotional styles. An emotional style, in his words, is:

> [A] consistent way of responding to the experiences of our lives. It is governed by specific, identifiable brain circuits and can be measured using objective laboratory methods... Because Emotional Styles are much closer to underlying brain systems than emotional states or traits, they can be considered the atoms of our emotional lives – their fundamental building blocks.

Emotional styles mark specific neural pathways that go beyond rudimentary explanations and discussions of emotions so often heard in our everyday lives, within and outside organizations. Based on rigorous and long-term scientific observations, Davidson established six emotional styles. Each style is a continuum between two extremes. We believe that it is vital for leaders to understand these in order to direct their behaviour better. Below is an explanation of each style and its relation to leadership, as we have observed and used it in our practice.

Style 1: Resilience

As mentioned in Chapter 1, on one extreme, this style is fast recovery from adversity and, on the other, slow recovery. It determines the way we respond emotionally to a negative event. If you cannot recover easily from a setback and you keep dwelling in negative emotions then you are closer to the slow recovery extreme. If your ability to function properly goes undeterred and you fight off setbacks quickly then you are closer to the fast recovery extreme. These two extreme emotional states depend on the interplay between the amygdala and the prefrontal cortex that we mentioned in Chapter 2. The more the amygdala is activated over the prefrontal cortex, the more towards the slow recovery extreme you will be. The more the prefrontal cortex is activated over the amygdala, the faster you will recover. Having in mind the increased dynamism of modern organizations and the 'small bets' approached discussed in Chapter 3, leaders need almost daily to move on quickly from disappointments and consider new alternatives, using high levels of energy. Resilience is, indeed, a key leadership characteristic that allows leaders to find the inner strength to lead themselves and their teams confidently and persuasively to new victories. We recommend that our students and organizational clients act quickly as soon as a setback occurs by looking for lessons learned, new alternatives and reasons to be active.

Style 2: Outlook

This is the popular pessimism–optimism continuum. As with Elaine Fox's dual system of rainy and sunny brain, outlook concerns the way by which we view everyday events and how much are we able to sustain positive feelings or not. Do we tend to search for the negative side of every situation, or do we 'always look on the bright side of life'? The interplay of the left prefrontal cortex and the nucleus accumbens, our pleasure centre, is in focus here. The more signals go from the prefrontal cortex to the nucleus accumbens, gearing it towards increased activity, the more you are on the positive extreme. The fewer signals arrive at the pleasure centre, the closer you are to the negative extreme. Our pleasure centre operates mainly under dopamine and opiates, and these two are responsible for different pleasure outcomes. Dopamine activity is associated with anticipation and opiates with pleasure. This means that leaders need to manage those two positive states differently. The feeling of excitement coming from anticipating a result, an event, a meeting, a presentation, is not based on the same chemicals responsible for the feeling of excitement when we actually achieve something. Dopamine is involved in the first, opiates in the second. A positive outlook is associated with higher motivation, energy, creativity, determination and health. We help the people we work for and with to move towards a more positive outlook by opiates enhancing both anticipatory and achievement emotions. Working towards a presentation in a great teamwork atmosphere and boosting morale by focusing on the upcoming experience and positive result can build dopamine-infused anticipation. Complementing everyone on their efforts, celebrating successes (even slight ones) and giving appropriate team and individual compensation can build opiate-infused reward (Le Merrer *et al*, 2009).

Style 3: Social intuition

Effectively reading the intentions and emotions of others is essential for great leadership, as no one can inspire others whilst being oblivious to their states of mind. On the one extreme of this continuum is a socially confused person and on the other is a socially intuitive person. When unable to decode someone else's emotions, we remain in the dark concerning possible courses of action to amend a situation and boost performance. However, when our sensitivity to other people's emotions is heightened we have the basis of empathy and compassion. Then, we are better fitted to respond appropriately and realistically to a situation. The amygdala is also at play here but this time together with fusiform gyrus. Fusiform gyrus is located on the temporal and occipital lobes of our cortex and has been associated with various types of recognition. A study published by Josef Parvizi of Stanford School of Medicine and his associates in 2012 proved, through the use of electrodes on a patient's brain, the sole responsibility of this brain region for facial recognition. When the amygdala is activated more than the fusiform gyrus when looking at people's faces, it is a strong indication of a socially confused style. When the opposite happens, this indicates a highly socially intuitive person. Dr Paul Ekman's (2007) pioneering work on studying human facial expressions all around the world has revealed the evolutionary role of our ability to identify other people's emotional state by observing their faces. Early in our development as a species, when our capacity for language was still limited, we depended on reading people's faces quickly and in a subconscious way in order to adjust our own behaviour and increase our chances of survival. To be socially aware, which is to develop social cognition, we are assisted not only by our evolved capacity to read faces. We automatically read eye movements, too. Humans have, uniquely, a visible large white part of our eyes in order to enable social communication through joint attention and signalling of our eye direction. The white part of our eyes makes it easier than in other species to trace where the black part is looking, and thus to infer intentionality. When the whole visible eye is black it is nearly impossible to make any assumptions about attention and preference. This cooperative eye hypothesis, as it is called, has been found to work well on both children and adults with a typical brain development, while not so well for people with autism, proving that successful social intuition requires the correct reading of the gazes of others (Segal *et al*, 2016). The innate ability to read faces and eyes has not disappeared because of our more advanced linguistic skills, but it is also not the same in all people, as mentioned earlier. To improve on this continuum we need to work both on quieting the amygdala, fighting its hijacking potential, and trying to become better at reading other people's emotional cues in their face, bodily movements, voice and actions. Observing your friends and close colleagues for emotional cues and discussing with them their emotional state by giving your evaluation is a good start. However, it has to be done with people who are close to each other and trusted in order to avoid any risks.

Style 4: Self-awareness

Are you able to determine your own emotional state accurately, or you are unable to detect or decode your emotions appropriately? Self-awareness is an emotional style with paramount importance to leadership since we can only direct our energy and actions appropriately when we know our thoughts and feelings by detecting accurately the messages our own bodies are sending us. For example, we should understand the impact of a difficult meeting on our emotional state and stop it from spilling over to subsequent unconnected meetings we might have. The main brain region responsible for this awareness is the insula, or insular cortex, which is part of our cerebral cortex. This brain region is related to consciousness and it is increasingly considered as the centre of our sentient self, or the subjective conscious self (see, for example, Craig, 2010 and Gu *et al*, 2013). This is because the insula contains a map of brain organs, receiving and sending signals to them. The higher the activity in the insula, the better the self-awareness, while the lower the activity in the insula, the worse the conscious experience and understanding of our own emotions. The famous ancient phrase 'Know thyself' inscribed on the Temple of Apollo in Delphi and attributed to Socrates, Plato and Thales among others, is the essence of the self-awareness continuum. On the one extreme, our ability for introspection is very weak, severing the connection between our inner self and our conscious self. On the other extreme, we can be very sensitive to our internal changes and signals. In affective neuroscience this introspection is called *interoception* and it refers to our capacity to identify and track internal bodily cues. This capacity, which has been traditionally linked to good physical and mental health and more recently to a healthy social life too, is composed of three major brain regions: the anterior insula cortex, the anterior cingulate cortex and the orbitofrontal cortex (Oldroyd *et al*, 2019). What the brain tries to do is to engage its internal radar, our nervous system, to detect changes in our body consciously and unconsciously in order to use these changes as crucial information for taking a decision or for fine-tuning an already existing one. The ultimate goal is to reach and retain a desired state of homeostasis that leads to success. This homeostasis, as explained by star neuroscientist Antonio Damasio, is not the same as the 'conventional and impoverished conception of homeostasis that confines itself to the "balanced" regulation of life's operation' (Damasio, 2018). This is a simplistic view. Rather, he continues, interoceptive-driven homeostasis:

> [E]nsures that life is regulated within a range that is not just compatible with survival but also conducive to flourishing, to a protection of life into the future of an organism or a species... It is my view that the unshakable imperative of homeostasis has been the pervasive governor of life in all its guises. Homeostasis has been the basis for the value behind natural selection, which in turn favors the genes – and consequently the kind of organisms – that exhibit the most innovative and efficient homeostasis.

Identifying internal body data and utilizing them to help us achieve the level of 'normality' our brain desires are key processes for brain-adaptive leadership. Even more important, of course, is the nature of this 'normality'. Is your brain happy with just having a job and thus trying to retain this positive homeostatic state by listening to feelings adjacent to this goal? Or is your brain happy with higher achievement and wider impact, and thus prioritizing feelings to act upon that are appropriate to such goals? Those brains with homeostatic states that are more innovative and efficient will survive longer and will be more evolutionarily successful. Leaders need to be particularly careful when recruiting for a position and when choosing members for a specific team since the homeostatic nature of the candidate should match that needed for the job and/or for the team. Unless, of course, you want to bring change into a team, and you opt deliberately for someone with a different homeostasis to other team members.

Our work with managers and leaders in organizations often includes long sessions discussing levels of awareness of own emotions in various work situations and their impact on overall behaviour. Taking time to quietly, and in isolation, assess your bodily signals when feeling excited, positively or negatively, can be crucial for understanding your emotional state accurately and acting accordingly. In our experience, self-awareness and purpose are closely related.

DECODING OUR EMOTIONS

Understanding our emotions is crucial for determining our emotional styles. However this is a very difficult task because of the difference in emotions and feelings. The two are not the same. Emotions are what is happening within us, whilst feelings are the *subjective perception* of what is happening within us and its elaboration. According to Damasio (2004), emotions are automated biological reactions in the form of chemical and neural responses to bodily or environmental stimuli. They are generated automatically, regardless of whether the stimulus is processed consciously or unconsciously, and their role is to make us move/react in order to survive. Feelings, on the other hand, are cognitive representations or perceptions of the changes caused by emotions as well as the evocation of thinking processes and mental states consistent with the emotion. If the insula works well then the awareness of emotions and the generation of corresponding feelings are congruent. We score high on the self-aware dimension. But what happens if those two are not consistent and we misinterpret the emotion for something that it is not, regardless of the working of the insula? Can the perception of our emotions be tricked by environmental and other conditions? And if yes, what can we do about it?

In a 1974 publication, psychologists Dutton and Aaron described their now legendary Capilano Suspension Bridge experiment. In the experiment, an attractive female researcher conducting a survey gave young male participants her phone number as part of the process, to answer to any additional questions after the experiment has taken place. The researchers measured how many of them called her back after the end of the experiment. One group of young men met the female researcher in the middle of an unstable suspension bridge and the other on stable ground. The results were clear: those who were on the suspension bridge contacted the female researcher more than those who were on safe ground. They were fooled by their own perception. The group on the dangerous bridge misattributed their feeling of arousal/excitement to the sexual attractiveness of the female researcher, rather than to their fear of survival. The issue is that the physical sensations of the two within our bodies are similar, and an external stimulus nudged them towards the wrong direction. A great and graphic case of inconsistent feelings and emotions.

Our recommendation for improving the perception of your own emotions, especially in professional situations, is the 3D+3L approach. This includes three types of descriptions and three types of listening:

1D. *Description of the situation.* Take your time and describe in detail a situation that made you experience intense feelings. Pay attention to causes, other people's roles and the physical surroundings. What forced you to feel like that? Consider this to be the equivalent of strategic environmental analysis. This description needs to happen relatively close to when the event occurred.

2D. *Description of reactions.* Continue with the detailed account of the behavioural response that the feeling produced, first in you and then to the others around you (if applicable). Track, in a step-by-step way, everything that followed, from having the feeling to playing it out. Write down both seemingly unimportant actions that only you know about, and significant ones that involved others. Why did you behave that way? This also needs to be done close to the event.

3D. *Description of correlations.* Create a list of events that had similar characteristics or similar behavioural outcomes. Are there any connections? Also, try to link your chains of thought to your feelings, the situation and the behaviours that followed. Are there any patterns? This is a more introspective analysis that is usually ongoing and generally takes more time than the previous ones.

1L. *Internal listening.* Look inside yourself with minimum interruption and with genuine interest to help you establish a stronger link between emotions and feelings. Silencing our typical chatter and ensnaring our sensitivity to raw emotions before we jump to easy conclusions are necessary steps in improving our understanding of our emotions. Mindfulness has become a global trend for CEOs and executives because of its promise of increasing self-awareness, as explained by a pioneer in the field, Maria Gonzales (2012). Give yourself the chances, planned and unplanned, to listen to your inner self.

2L. *External listening.* Listen to others carefully and try to understand their viewpoint, without letting your strong feelings interfere with/misinterpret the conversation. This is key to having a clear emotional picture of the situation. Furthermore, asking others to evaluate and give you feedback on your emotional state can help bypass our internal miscommunication of our own feelings. Don't hesitate to ask the ones closer to you.

3L. *Constant listening.* Developing a mental radar that constantly scans internally and externally for emotional cues and their impact on you and those around you is the ultimate goal for self-awareness. Your own personal 'emotions observatory' should be established in your mind that will relentlessly evaluate important and less important moments for emotional reactions and that will set a knowledge directory for emotional patterns and related behaviours.

ACTION BOX: THE 3D+3L APPROACH IN PRACTICE

Think of a recent emotionally challenging situation that you have passed through with more or less success. Then, use the three types of descriptions and three types of listening above in order to write down how this situation could be interpreted. What are your conclusions? Are there things that you could do differently regarding this situation?

Style 5: Context sensitivity

Are you emotionally harmonized with what is happening around you, and are you behaving accordingly? Or are you usually caught unawares by the situation, adopting inappropriate behaviour that sometimes makes you embarrassed? At one extreme in this continuum is people who are 'tuned in' to the environment, and at the other 'tuned out'. Tuned in means emotional synchronization while tuned out means emotional disconnection with a situation. We have seen, for example, managers making jokes at a crucial moment of a presentation, or being over-serious when a business partner tries, correctly, to lighten up a conversation. The brain region associated with such emotional synchronization is the hippocampus. We have already discussed the hippocampus in relation to memory formation, but it plays also a vital role in the fitting in or tuning behaviours related to contexts or situations. A weaker or smaller hippocampus can result in completely inappropriate emotions, and subsequent behaviours, in response to external conditions. Furthermore, contextual learning is diminished in such cases. Thus we need to make sure that our internal state and external requirements fit both in order to behave accordingly and to take away the right lessons from the situation.

The highest level of tuning in is when our brain is fully connected, or 'coupled', with the brain of the person we try to communicate with. This brain coupling occurs when the two brains go beyond just observing, processing and mirroring each other's concrete words and actions, and move into more complex connectivity concerning meaning and intentionality (Hasson and Frith, 2016). In advanced tuning, the same brain areas fire between people: not just those involved in the simple perception of movement and speech but the ones that create the deep sense of a dynamic and shared social reality that 'takes two to tango'. This is what happens when we see people in the workplace collaborating eloquently and effortlessly, when they seem to read each other thoughts in meetings and when they actively help each other even before they are asked. A study by Jiang *et al* (2015), based on hyperscanning (fNIRS) experiments, shows that there is higher neural synchronization between leaders and their followers in comparison to leaderless dyads. The result of this study supports the view that a strong relationship between the leader and the follower is very important for the effectiveness of the leadership process. A strong relationship is enhanced by high brain connectivity. Brain connectivity can be enhanced through high quality of communications between the leader and the follower.

Some of the recommendations made in the 'Decoding our emotions' box apply here, too. Listening to others and creating as accurate a description as possible of what happened can lead us to pattern recognition, learning and appropriate behavioural adaptations in the future. A note of caution: if you are not sure if what you sense in the meeting room is tension or positive anticipation then it is better to wait for more cues until you make a decisive move. If you feel lost constantly, then focus on working on moving up the continuum and towards a more tuned in style. But be careful – too much sensitivity can increase stress responses, since it is associated with anxiety and feelings of inadequacy. As Ellis and Boyce (2008) have argued, although sensitivity to context in our brains is evolutionarily invaluable for learning to deal effectively with the environments we find ourselves in, it follows a U-curve in relation to stress, meaning that low sensitivity produces high stress, as does high sensitivity. Being a little disconnected, or not fully connected, is a great asset for modern leaders in our unpredictable and stressful modern business environment. How disconnected is something that each leader needs to determine individually based on his/her experience and personality as well as on the situation.

Style 6: Attention

On the one extreme of this dimension is a focused person; while on the other extreme is an unfocused one. Do you have trouble focusing on the person talking to you from across the meeting table? Do you feel your mind wandering every time you look at the detailed monthly results or the latest employee engagement scores from HR's research? Are you often annoyed at your open-plan office because the noise does not allow you to do your job properly? Then you are closer to the unfocused end. In such cases, your emotional awareness will be limited and your behavioural response unsuitable to the situation at hand. The prefrontal cortex, or executive brain, is responsible for focus since higher activity can produce a condition called 'phase locking', which is the ultimate focus. In this condition, the engagement of the prefrontal cortex is perfectly timed with the external stimulus, meaning that you are mentally in line with what you are paying attention to. This kind of 'phase locking' is much admired in corporations. A Bloomberg article authored by Buhayar (2013) reported on the popularity in CEOs' speeches of the phrase *laser focus*. Being laser focused on execution, growth, new products and market opportunities among others has apparently entered the jargon of top-level communications making sure that distractions will not be tolerated. However, too much focus is not always an advantage. A study by neuroscientist Jason Gallate and his associates (2012) revealed that creative people benefit the most when taking a break from focusing on a given problem and that this break allows for non-conscious processes to take over and provide insightful solutions. Too much focus inhibits creativity, but loose focus inhibits performance and productivity. At the same time, there are situations when extreme focus is needed, like in certain negotiations, and thus the intensity of focus has to be determined on a case-by-case basis. Having a strong purpose, as explained in Chapter 1 and applying the 'small bets' method explained in Chapter 3 still leaves room for a more chaotic, complex and seemingly unfocused approach in goal-setting and decision. As economist John Kay explained after examining diverse examples of goal setting and achievement (*Independent*, 2010):

> Problem-solving is iterative and adaptive, rather than direct. Good decision-makers are balancing incompatible and incommensurable objectives. They are eclectic and tend to regard consistency as a make of stubbornness, or ideological blindness, rather than a virtue.

So, we should obsess less with setting clear and consistent goals in our life and business. Instead, we should search for the surprising success of intuitively following unbalanced and disordered objectives that, at the end, are more successfully achieved. The hidden costs of always seeking clarity and focus in decision-making are higher than we might think.

Continuous practice is the key to moving to one or the other extreme in all of the styles above, making use at the same time of something we referred to in Chapter 3 as neuroplasticity: the ability to change our neural pathways and thus change positions on the emotional styles' continuums. Although it might be challenging to go from one extreme to the other, moving along the line is both possible and advisable. We have seen it happening both in ourselves and in the people we have been working with. Putting this as a regular practice in your agenda will provide more successful results.

ACTION BOX: MEASURE EMOTIONAL STYLES IN TEAMS

Go through all six dimensions by Davidson and score yourself from 1 to 5, 1 being the low extreme and 5 being the high extreme. Reflect on every score. Write down a few bullet points on the scores you want to retain and the ones you want to change, upwards or downwards. Ask your closest associates to score you on all six and observe any deviations between your scores and those your associates gave you. Why do you think they exist, if they do? You can do this exercise as a team as well, each one first scoring themselves and then scoring each other. An open discussion based on trust and positive feedback can do a lot to help each other improve on the dimensions. This is not to be tried initially in the team setting without first having professional and experienced help.

From mood to great feelings

The long-term effect of emotional styles on how we think and behave in our organizations, although not the same, resemble the long-term effects of personality: they follow us through a large part of our lives. They can change to better fit our leadership challenges but this takes time and much targeted effort. In a more short-term horizon, our emotions are highly influenced by our moods, and modern leaders need to be aware of them and try to control them, since they form the fertile ground for our everyday emotional reactions. Although themselves hidden, they are responsible for most of the emotions we experience in our workplace. According to one of the leading experts on moods, Dr Liz Miller (2009):

> We all have moods all the time... Moods are an internal measure of how we are. We do not express our moods directly. Instead we express them indirectly in the way we think, communicate, behave and see the world... Almost all anger reflects an underlying irritable, anxious mood. This mood has provided the soil that allowed the emotion of anger to grow.

It is your general mood that will determine greatly the types of emotion you will express at any given moment. Thus, by getting into the right mood and avoiding counterproductive ones, you can reach for the right set of emotions in any situation and use them to promote further your leadership abilities. For example, how much better can you manage your emotions and valuable contribution in a difficult meeting about the latest increase in employee absenteeism if you are in a positive mood compared with a negative one? We have encountered countless managers that have confided in us that they wished they had reacted differently in a stressful situation, but that the overall bad mood in the company 'during those days' did not allow them to do so. If they knew how to set their mood to positive they would have dealt much better with any challenging situation. This is because, in the long list of benefits, nurturing a positive mood is integral to concentration and clear thinking, managing your behaviour, creating meaningful and lasting relations, flexibility and perseverance in pursuing goals and work satisfaction. In a particularly interesting study, Ruby Nadler of the Western University Canada and her colleagues (2010) have demonstrated that people in a positive mood outperform those in neutral or negative moods. Those in positive moods show higher cognitive flexibility and do much better in tasks that require heavy use of the prefrontal cortex, our executive brain, such as hypothesis testing and rule selection. This has a profound effect on management, since a positive mood emerges as a prerequisite not only for the softer aspect of the job, such as keeping the team up-beat and motivating individuals, but mainly for better utilizing our cognitive processing capacities. That is, to think better and come up with better solutions in difficult problems. Good mood means good thinking.

Miller's (2009) now famous categorization of moods uses two variables to plot a 'mood map': energy on one axis and well-being on the other. The first axis (vertical) has high energy on the top and low energy on the bottom. The second (horizontal) has negative well-being on the left and positive on the right. Energy relates to the neurotransmitter dopamine, which is about arousal/excitement; and well-being relates to the serotonin–endorphin system, which is about positive feeling, good functioning of internal organs and pain relieving. The four quadrants of the mood map and the feelings belonging in each one are explained below, following our practice with managers worldwide:

Q1. *The rocket mood.* High energy, positive well-being quadrant. Strongly motivated, highly pleased and continuously excited leaders that drive everyone forward with passion and conviction. Fast-moving, with unlimited energy and a great smile on their face, leaders in the rocket mood are the guiding and unstoppable force of their teams. This mood rarely gets you tired and when it does you bounce back with even more resourceful ideas and precious emotional support for others.

Q2. *The guru mood.* Low energy, positive well-being quadrant. Calm, highly content and satisfied leaders that create the conditions for thoughtful work, a peaceful office environment and the unbeatable feeling of safety. Leaders in the guru mood are an effective weapon against the craziness of a situation and create the necessary stability for reaching wise decisions with high long-term effects.

Q3. *The downer mood.* Low energy, negative well-being quadrant. Physically and mentally exhausted, often sad and even depressed leaders that cannot hide their boredom alongside their negative feelings at work. Leaders in the downer mood affect negatively all around them, sucking in energy and diffusing pessimism.

Q4. *The panic mood.* High energy, negative well-being quadrant. Tension, irritation, fear and frustration are a few of the powerful but negative feelings expressed by leaders in this mood. A jumpy attitude, a strong focus on self-preservation and the image of 'running around like a headless chicken' are associated with leaders in the panic mood that seem to make everyone around them nervous and insecure. In Table 4.1 you can see the mood map in a matrix table.

TABLE 4.1   Mood map matrix

	Well-being	
Energy	*Negative*	*Positive*
High	The panic mood	The rocket mood
Low	The downer mood	The guru mood

A leader does not have to be on the extremes though to identify his/her mood as *rocket*, *guru*, *downer* or *panic*. Even milder versions capture the essence of each category. Keeping yourself on the positive side and increasingly on the high energy side is of paramount importance. In fact, Alison

Wood Brooks (2013) from Harvard Business School found that when there is emotional arousal because of an upcoming important challenge, like a presentation or a test, people do much better when they interpret this mood as excitement than anxiety. And a positive, rocket mood helps you do exactly that. She also found that people trying to use the old strategy of calming down, do badly. Another nail on the coffin of the 'staying cool' approach.

Miller's model takes inspiration from the older model of categorizing emotions called Russell's circumplex model of affect. This model is particularly helpful since it deals effectively with many misconceptions about emotions. For example (Posner *et al*, 2005):

> The role of the amygdala in the processing of emotional stimuli has proved elusive, with early neuroimaging studies correlating amygdala activation with fear responses and other negative emotions, but underestimating the role of the amygdala in subserving positive emotions. The circumplex model suggests that evidence of amygdala activity during both aversive and appetitive [see approach and avoidance neural networks] stimuli can be understood readily if the amygdala is a part of the arousal system... Neuroimaging paradigms exploring olfactory and gustatory stimuli support this view by demonstrating that amygdala activity increases with arousal, regardless of the particular positive or negative valence associated with experience of the stimuli.

This means that amygdala is not associated exclusively with negativity. Its role is more complex and interesting since it has been found to activate in both positive and negative arousal states. Actually, when older generations of teachers and managers alike were calling for their students and employees respectively not to be 'overly emotional', they meant 'emotionally aroused'. Emotions are always present but their arousal level and their valence, which is their positive or negative nature, change depending on the brain's motivation and intentionality.

ACTION BOX: CHECK YOUR MOOD.

Mentally revisit the last two weeks at work. Did you have a prevailing mood throughout or did you experience different moods at different times? Write down the mood(s) you recognize and try to categorize it/them using the model explained in this part of the book. By using the four quadrants of the mood map, try to describe your experience based on Q1, Q2, Q3 or Q4. In order to confirm your conclusions it is always a good strategy to ask the people that you trust and interact with more, especially those close to you in the period of time that you evaluate. Write down the causes and outcomes. Repeat the exercise once every month. After a few months try to identify patterns based on the following questions:

- Are specific moods appearing because of specific causes, having specific outcomes?
- Why do you think this is the case?

Since the link between positive moods and better performance is scientifically established, your goal is to move towards sustained positive moods like the rocket and the guru. Achieving a positive mood requires changes in the environment, in awareness/knowledge, in relationships and in emotional styles as described in the previous section. Do not allow your mood to damage your leadership potential. On the contrary, put your mood to work for you and your emotions. Proactively, and after you become comfortable in detecting your mood, try to do so before important meetings and presentations, and go in with the most appropriate one.

Moods are not as fixed as emotional styles and personalities, but they can set themselves in for a long time if not challenged. Having a constantly negative mood is a sign of the need for change, and leaders should often consider their mood to make sure they give themselves the chance to experience the right emotion at the right time.

The importance of having the right mood at work is underlined by the fact that our emotions are not only within us but 'spill over' to those around us. The phenomenon of emotions spreading like a virus from one person to another, or from a person to a whole team, is called emotional contagion. As Snaebjornsson and Vaiciukynaite (2016) found, early psychology and organizational studies have recognized that emotions and moods spread between individuals and that this has a considerable impact on performance. Leaders do indeed affect people around them with their emotions, whether or not those people are conscious or aware of this process. The process of emotional contagion includes the mechanisms of mimicry, brain synchronicity/coupling, emotional shared experience, and conscious and subconscious emotional feedback (Tee, 2015). Below are the key insights for emotional contagion and leadership (Barsade *et al*, 2018):

- Emotional contagion influences team performance. Leaders radiating a positive mood enable teams to perform tasks faster, with increased coordination and with heightened creative problem-solving abilities.

- Leaders that manage to spread their emotions better, especially positive ones, are perceived as more charismatic by people and thus leadership effectiveness increases.

- Emotional contagion depends both on the leaders' ability to transfer emotions and the openness of people to receive them.

- Leaders might be effective in transmitting their emotions to team members, but team members must then circulate the same emotions between themselves.

- Emotional contagion is a two-way phenomenon. This means that leaders are more successful in transferring emotions when they themselves are more open to receive emotional contagion from their teams.

- Emotions are transferred unconsciously but they can also be transferred consciously, especially when they are authentic and transparent. People intuitively, rather than analytically, feel that this phenomenon exists and that it's potentially very powerful.

Adopting the right mood, or long-term emotional state, is crucial for leaders, not just for their personal performance but primarily for the performance of others through the remarkable process of emotional contagion.

EQ as an empowerment qualification

Since the mid 1990 when Daniel Goleman (1995) spread the word for the merits of emotional intelligence, or emotional quotient (EQ), with his international best-selling book, there have been countless other books, white papers, academic research papers, consulting and training events, services and experts on the topic. Multinational corporations have adopted the practice of evaluating their executives' emotional intelligence, and the concept has even been found to help managers fulfil their leadership potential since, as Dijk and Freedman (2007) illustrated, employees become more skilful in emotional literacy and subsequent thinking as they climb up the organizational ladder. For us, EQ was seminal – and still is – in changing the outdated mindset in societies and businesses that emotions are inappropriate and even destructive. As excellently put by Cooper and Sawaf (1996) in their early and highly influential book *Executive EQ*, we need to change our view of emotions:

- from an interference with good judgement to essential to good judgement;
- from a force of distraction to a force of motivation;
- from a sign of vulnerability to a sign of vitality and presence;
- from an obstruction to reasoning to an enhancement and acceleration to reasoning;
- from a barrier to control to the foundation of trust and connection;
- from an inhibition to objective data to a significant information and feedback mechanism;
- from a complication to planning to the main spark of creativity and innovation;
- from weakening our attitudes to activating our morality;
- from undermining our authority to boosting our influence without authority.

We have to attribute to the EQ movement the wave of change in organizations worldwide of accepting emotions as a reality that needs to be incorporated into both mindsets and processes. Boyatzis and Goleman (1996), in their book *Emotional Competency Inventory*, suggest an emotional intelligence model for leaders developed as the first holistic approach in managing emotions that inspired the masses and made its way so profoundly into corporations, big and small. Both scholars and their associates expanded and developed this approach further in later works (Goleman *et al*, 2002; Boyatzis and McKee, 2005). This holistic emotional intelligence model, which we also continue to practice in our consulting, coaching and training engagements, consists of four major clusters of which two concern our internal world and two the external:

Cluster 1: Self-awareness (internal awareness) is about being able to recognize our own emotions and consists of:
- o emotional self-awareness: recognizing our emotions and their effects;
- o accurate self-assessment: knowing one's strengths and limits;
- o self-confidence: a strong sense of one's self-worth and capabilities.

Cluster 2: Self-management (internal management) is about being able to change our own emotions and consists of:

o   adaptability: flexibility in dealing with changing situations or obstacles;

o   self-control: inhibiting emotions in the service of group or organizational norms;

o   optimism: a positive view of life;

o   initiative: proactive bias toward action;

o   achievement orientation: striving to do better;

o   trustworthiness: integrity or consistency with one's values, emotions and behaviour.

Cluster 3: Social competence (external awareness) is about being able to recognize emotions in others and consists of:

o   empathy: understanding others and taking an active interest in their concern;

o   service orientation: recognizing and meeting customers' needs;

o   organizational awareness: perceiving political relationships within the organization.

Cluster 4: Relationship management (external management) is about being able to change emotions in others and consists of:

o   inspirational leadership: inspiring and guiding groups and people;

o   developing others: helping others improve performance;

o   change catalyst: initiating or managing change;

o   conflict management: resolving disagreements;

o   influence: getting others to agree with you;

o   teamwork: creating a shared vision and synergy in team work, collaborating with others, building relationships and networks.

In a recent study, Eurich (2018) found that internal self-awareness is related to better job satisfaction, self-control, better relationships and happiness, and also negatively related to stress and anxiety. External self-awareness is associated with more empathy and taking into considerations others' perspective, which is even more critical nowadays. For leaders it is critically important to develop both types of self-awareness in order to develop better relationships (see also Chapters 7 and 8) with followers.

In a nutshell, EQ helps leaders become aware and manage their own emotions and the emotions of others. It starts from inside you and spreads out to the people around you. Empathy, active listening and intrinsic motivation are only a few of the keywords that EQ spearheaded bringing into our lives and they seem to be key elements in leadership rhetoric. As Antonacopoulou and Gabriel (2001) argue, these elements are necessary in leading with care. They also suggest that a leader needs to have 'guts' in order to lead. Having the guts to lead means emphasizing the following three emotional dimensions:

- emotions as coping mechanisms (emotions can help us to adapt to changing circumstances);
- emotions as transitional qualities (emotions preserve what a person values in different circumstances);
- emotion as a system of reactions (emotions support the interpretation of situations).

In other words, emotions are key elements in engaging leadership. In our experience, EQ is a true empowerment qualification since without EQ a leader is unable to navigate through and appropriately manage emotional styles and moods internally and externally. Any capacity to inspire, influence and move others is minimized and the person will soon feel outworked, outperformed and eventually like an outcast. On the contrary, EQ brings an ideal predisposition for working with emotions and a high probability for succeeding in the long run. This is the reason why we consider EQ as emotional competence that can be developed and without any doubt can help leaders improve their effectiveness and improve the overall relationship between leaders and their followers.

Regardless of the widespread appeal of EQ and its two decades of influence, some managerial and organizational systems are yet to take into account fully the teachings and practices of EQ. There are training and other initiatives that have moved in this direction, but overall organizations still look and feel emotions-averse more than emotions-run. Since our brain uses emotions in such diverse ways and uses them to boost both thinking and performance we hope that a new wave of emotional intelligence will hit companies. This new wave is the neuroscience one. All in all, we cannot be too overemotional about the importance of emotions in leadership.

Keep in mind

A brain without emotions is a malfunctioning brain, and this is something that organizations, leaders and managers have to understand quickly in order to calibrate better towards a brain-based leadership approach. An important aspect towards BAL is homeostasis, or a balanced regulation of our life. In addition, brain coupling is critical for collaboration and collective performance. A common denominator of homeostasis and brain coupling is our emotions. Emotions help us think faster and morally, move others and ourselves, and boost our cognitive capacity for advanced performance. The latter can be achieved through emotional contagion. Moreover, emotional intelligence is a critical leadership capability. Being aware of your emotional styles and moods, and applying a holistic approach that takes into account both the internal and external worlds, can help you bypass the outdated notion of incompatibility of emotions and business. Leadership is primarily an emotional skill and CEO truly means chief emotional officer in any type of organization.

BOOST YOUR BRAIN: IDENTIFY EMOTIONS IN LEADERSHIP

Think of a leader for whom or with whom you have worked a lot, and especially someone that you would gladly work with or for again. Then, think of a person in a leadership position that you either try to avoid, or who left little impact on you. Then try to compare both people by writing down how each of them acted in their daily professional lives. Also, think and write down how they related to others. What lessons can be learned from this comparison? Can you identify the power of emotions? Consider your conclusions and use them next time.

References

Antonacopoulou, EP and Gabriel, Y (2001) Emotion, learning and organizational change: Towards an integration of psychoanalytic and other perspectives, *Journal of Organizational Change Management*, 14(5), pp 435–51

Australian Psychological Society (2016) Corporate psychopaths common and can wreak havoc in business, researcher says, www.psychology.org.au/news/

media_releases/13September2016/Brooks/ (archived at https://perma.cc/ Q5YX-MNXP)

Babiak, P and Hare, DR (2006) *Snakes in Suits: When psychopaths go to work*, HarperBusiness, London

Barsade, SG, Coutifaris, CG and Pillemer, J (2018) Emotional contagion in organizational life, *Research in Organizational Behavior*, 38, pp 137–51

Boyatzis, R and Goleman, D (1996) *Emotional Competency Inventory*, The Hay Group, Boston, MA

Boyatzis, R and McKee, A (2005) *Resonant Leadership: Sustaining yourself and connecting to others through mindfulness, hope, and compassion*, Harvard Business School Press, Boston

Brooks, AW (2013) Get excited: Reappraising pre-performance anxiety as excitement, *Journal of Experimental Psychology: General*, 143(3), pp 1144–58

Buhayar, N (2013) 'Laser-focused' CEOs proliferate as jargon infects speech, www.bloomberg.com/news/articles/2013-09-11/laser-focused-ceos-multiply-with-promises-from-ipads-to-macaroni (archived at https://perma.cc/QNR8-TH3E)

Chamorro-Premuzic, T (2019) 1 in 5 business leaders may have psychopathic tendencies – Here's why, according to a psychology professor, www.cnbc.com/ 2019/04/08/the-science-behind-why-so-many-successful-millionaires-are-psychopaths-and-why-it-doesnt-have-to-be-a-bad-thing.html (archived at https://perma.cc/7GUT-YSWZ)

Conley, C (2012) *Emotional Equations: Simple truths for creating happiness + success*, Simon and Schuster, New York

Cooper, RK and Sawaf, A (1996) *Executive EQ: Emotional intelligence in leadership and organizations*, Perigee, New York

Craig, AD (2010) The sentient self, *Brain Structure and Function*, 214(5–6), pp 563–77

Damasio, A (1994) *Descartes' Error: Emotion, reason and the human brain*, Penguin, New York

Damasio, A (2004) Emotions and feelings: A neurobiological perspective, in *Feelings and Emotions: The Amsterdam symposium*, ed AS Manstead, N Frijda and A Fischer, pp 49–57, Cambridge University Press, Cambridge

Damasio, A (2018) *The Strange Order of Things: Life, feeling and the making of cultures*, Pantheon Books, New York

Davidson, JR and Begley, S (2012) *The Emotional Life of Your Brain: How to change the way you think, feel and live*, Hudson Street Press, New York

Dijk, CFV and Freedman, J (2007) Differentiating emotional intelligence in leadership, *Journal of Leadership Studies*, 1(2), pp 8–20

Dutton, DG and Aaron, AP (1974) Some evidence for heightened sexual attraction under conditions of high anxiety, *Journal of Personality and Social Psychology*, 30, pp 510–17

Ekman, P (2007) *Emotions Revealed: Recognizing faces and feelings to improve communication and emotional life*, Owl Books, New York

Ellis, BJ, and Boyce, WT (2008) Biological sensitivity to context, *Current Directions in Psychological Science*, **17**(3), pp 183–87

Eurich, T (2018) What self-awareness really is (and how to cultivate it), *Harvard Business Review*, January, p 4

Fayol, H (1930) *Industrial and General Administration*, Sir Isaac Pitman and Sons, London

Fayol, H (1949) *General and Industrial Management*, Sir Isaac Pitman and Sons, London

Fox, E (2013) *Rainy Brain, Sunny Brain: How to retrain your brain to overcome pessimism and achieve a more positive outlook*, Arrow Books, London

Galetzka, C (2017) The story so far: How embodied cognition advances our understanding of meaning-making, *Frontiers in Psychology*, **8**, p 1315

Goleman, D (1995) *Emotional Intelligence: Why it can matter more than IQ*, Bantam, New York

Goleman, D, Boyatzis, R and McKee, A (2002) *Primal Leadership: Realizing the power of emotional intelligence*, Harvard Business School Press, Boston

Gonzales, M (2012) *Mindful Leadership: The 9 ways to self-awareness, transforming yourself and inspiring others*, John Wiley and Sons, Chichester

Gu, X, Hof, PR, Friston, KJ and Fan, J (2013) Anterior insular cortex and emotional awareness, *Journal of Comparative Neurology*, **521**(15), pp 3371–88

Hasson, U, and Frith, CD (2016) Mirroring and beyond: Coupled dynamics as a generalized framework for modelling social interactions, *Philosophical Transactions of the Royal Society B: Biological Sciences*, **371**(1693)

Independent (2010) Think oblique: How our goals are best reached indirectly, www.independent.co.uk/arts-entertainment/books/features/think-oblique-how-our-goals-are-best-reached-indirectly-1922948.html?amp (archived at https://perma.cc/JNH8-K795)

Jiang, J, Chen, C, Dai, B, Shi, G, Ding, G, Liu, L and Lu, C (2015) Leader emergence through interpersonal neural synchronization, *Proceedings of the National Academy of Sciences*, **112**(14), pp 4274–79

Kaczmarek, LD, Behnke, M, Enko, J, Kosakowski, M, Guzik, P and Hughes, BM (2019) Splitting the affective atom: Divergence of valence and approach-avoidance motivation during a dynamic emotional experience, *Current Psychology*, 23 April, pp 1–12

Kelley, NJ, Hortensius, R, Schutter, DJ and Harmon-Jones, E (2017) The relationship of approach/avoidance motivation and asymmetric frontal cortical activity: A review of studies manipulating frontal asymmetry, *International Journal of Psychophysiology*, **119**, pp 19–30

Le Merrer, J, Becker, AJJ, Befort, K and Kieffer, LB (2009) Reward processing by the opioid system in the brain, *Physiological Reviews*, **89**(4), pp 1379–1412

Miller, L (2009) *Mood Mapping: Plot your way to emotional health and happiness*, Rodale, London

Miller, M, Bentsen, T, Clendenning, DD, Harris, S, Speert, D and Binder, C (2008) *Brain Facts: A primary on the brain and nervous system*, Society for Neuroscience, Washington

Nadler, RT, Rabi, R and Minda, JP (2010) Better mood and better performance: Learning rule-described categories is enhanced by positive mood, *Psychological Science*, **21**(12), pp 1770–76

Oldroyd, K, Pasupathi, M and Wainryb, C (2019) Social antecedents to the development of interoception: Attachment related processes are associated with interoception, *Frontiers in Psychology*, **10**, pp 712

Parvizi, J, Jacques, C, Foster, BL, Withoft, N, Rangarajan, V, Weiner, KS and Grill-Spector, K (2012) Electrical stimulation of human fusiform face-selective regions distorts face perception, *Journal of Neuroscience*, **32**(43): pp 14915–20

Posner, J, Russell, JA and Peterson, BS (2005) The circumplex model of affect: An integrative approach to affective neuroscience, cognitive development, and psychopathology, *Development and Psychopathology*, **17**(3), pp 715–34

Ronson, J (2011) *The Psychopath Test: A journey through the madness industry*, London: Picador

Segal, NL, Goetz, AT and Maldonado, AC (2016) Preferences for visible white sclera in adults, children and autism spectrum disorder children: Implications of the cooperative eye hypothesis, *Evolution and Human Behavior*, **37**(1), pp 35–39

Snaebjornsson, IM and Vaiciukynaite, E (2016) Emotion contagion in leadership: Follower-centric approach, *Business and Economic Horizons (BEH)*, **12**(1232-2017-2389), pp 53–62

Taylor, WF (1911) *The Principles of Scientific Management*, Harper and Brothers, New York and London

Tee, EY (2015) The emotional link: Leadership and the role of implicit and explicit emotional contagion processes across multiple organizational levels, *The Leadership Quarterly*, **26**(4), pp 654–70

Tomasello, M (2019) Michael Shermer with Dr Michael Tomasello: Becoming human, Science Salon 64, www.youtube.com/watch?v=ghJkITn-il8 (archived at https://perma.cc/XL6N-R5FV)

Weber, M (1947) *The Theory of Social and Economic Organization*, Free Press, New York

0 5

Right emotion, right action

FEAR EQUALS HARD WORK

No one should underestimate the power of fear in the workplace. Especially the executive of our real-life case who, during her early professional years, worked closely with a boss who led with an iron fist and always did her best to spread fear and terror throughout the office. Although disagreeing with such methods in principle, she instinctively learned from this boss that people will focus more, work harder, and stay longer out of fear rather than incentives. She just had to observe her own reaction to the emotional environment created by her boss to understand how effective fear was. Sure enough, some people could not handle it and either broke down or left (or both). But the necessity for such an approach was apparent in the challenges the company faced at the time: the fierce competition, political turmoil and high stakes were just too much for a fluffy, soft approach. This was a game for strong people and only the strongest would survive. And so she did.

Becoming a top executive in the same company took her several years. But years did not take away what she learned in those early days. Fear was her biggest weapon both for her own motivation and for her subordinates. A generally fearful culture in her department helped her to keep people committed, productive and more easily managed. Fear kept her in control, and control was all she needed to manage herself and others. She believed in a simple equation: fear leads to control, control leads to effective management, and effective management leads to great performance. However, she was not completely blind to the advantages of positive motivators. She used both informal and formal positive measures when needed to achieve particular objectives, like attracting new personnel, keeping her team from constant emotional collapse, preventing people from leaving the company and sometimes for feeling more humane herself. After all, she believed that humans were inherently good and kind and she was a great person to most of her friends and family. It is just that serious company work required a 'serious' approach. Thus, the stick was for keeping people in order and constantly running, and the carrot for attracting new ones and preventing the good ones from leaving. The 'carrot and stick', or better, the 'stick and the carrot' were her leadership toolbox. So simple, so effective.

However, she started facing problems when a new chief of human resources (CHR) pointed out in a high-profile meeting that they could not afford to have any more great talent fleeing the company. This was because, according to the latest corporate culture survey, the climate in many departments had to urgently change to become more positive. The department that scored worst on this internal research was hers. Also, her department had the highest rate of talent turnover. Following those numbers, the CEO requested a personal meeting with her. In that meeting she had the opportunity to explain her approach and how effective it had been up until now, and that she saw no reason to change her 'stick and carrot' approach. She claimed she knew how to use emotions to motivate her employees as well as herself and that she was upset about the new situation. The CEO, in cooperation with the new CHR, asked her to go through an intensive emotions training programme and assigned her a new professional coach, specializing in emotional management. After some time, she claimed her life had changed and that she now saw how beautifully complicated human emotions are. She was a much better manager and a great leader to her team and to her colleagues. It was now clear to her that serious business needs a serious approach to emotions, and not the simplistic one she was applying.

Why have management and business in general been allergic for so long to a serious and deep discussion of emotions? And why, when acknowledging that emotions might actually play a role, did they use humiliating concepts like the 'carrot and the stick', which is not even appropriate for donkeys, never mind humans? Fear, as a core management tool, is widespread in organizations of any type and size, as our experience around the world suggests. Simplistic views of emotions in companies are dominating management styles, partially because the neurobiological basis of emotions is missing from basic, and even from advanced in many cases, management education and training. In other words, the majority of management education programmes around the world, and especially MBAs, are based on principles and practices supported by the traditional jargon of management science, neglecting critical findings of other disciplines that specifically call for emotional and social intelligence to become integral skills in management and leadership (Heiss, 2014). What emotions do we have and how can leaders acknowledge them and use them within themselves and others in their organizations? Emotions have tremendous power to move people, and modern leaders can no longer be oblivious of how they actually work and of the combinations that can work better for them in their organizations.

The basic emotions in the brain

There is a wider consensus, as seen in the previous chapters, that emotions are necessary for human motivation and action. Without emotions, there is no urgency to behave in any way; to transfer dynamically from an undesired position of pain, loss and defence to a more desirable position of happiness, gain and protection. Since movement is the primary function of the brain (Wolpert, 2011), emotions, as the fuel for that movement, should be placed in the very centre of contemporary leadership approaches. Indeed, movement and brain are so tightly related that, as neuroscience researcher and author Baroness Susan Greenfield, proclaimed (2015):

> For stationary life forms, a brain is no longer necessary. The whole point is that for an animal moving around, there is an interaction with an environment that is incessantly changing. You need a device to tell you very quickly what is happening and, most importantly, to enable you to respond to what is happening, to get out of the way of predators or to chase after prey. So the brain, in whatever shape, size, and degree of sophistication, is somehow connected in a very basic way to ensuring survival as both a consequence and a cause of movement.

Trees and plants, although they do have diverse internal systems that allow them to survive and thrive, they do not possess a clear administrative centre that collects all sensory, internal and external, information to produce an integrated picture of a situation and to respond to that situation decisively and as fast as possible. Animals, on the other hand, and any multi-cell organism, are equipped with a CEO that sees, hears and feels everything in order to make decisions quickly and act accordingly. This CEO is the brain, and its main function is to initiate movement to ensure success. Thinking, feeling, understanding and any other brain processes are useful only in their capacity to assist the brain to move the body in the right direction. Although basic human drives like hunger, motivation, homeostasis and emotions are not exactly the same phenomena from a neuroscientific perspective (Adolphs and Anderson, 2018), there is a widespread consensus that all these, often collaboratively, lead to movement, which is what the brain is made for.

Emotions are the brain's integrative compass for action. Thus, the higher the need for change, even for constant transformation as in many contemporary industries, the more urgent the need for emotions to keep people changing, taking the initiative and bringing new solutions to new problems. The brain is actually built for that, but are we using it appropriately?

The emotional state in the brain is often described as a continuum, with positive and negative on the two extremes and neutral in the middle. Both extremes are there to make us move, either towards or away from an object, a person or a situation. As described in the previous chapter, these two conditions are the 'avoid vs approach' motivational systems in our brains in which, depending on the decoding of the information reaching the brain, different circuits are activated to make us act accordingly. The important point here, though, is that the number of emotions that exist on this continuum and form the toolbox for leaders to use on themselves and others to activate avoid/approach systems is rather established in psychology and neuroscience. We will examine a few of them here, highlighting both similarities and differences in order to demonstrate how they work.

We have previously separated emotional states (emotions) from their subjective perception (feelings), their mid-term patterns (moods) and their long-term, personality-like models (emotional styles), and analysed moods and styles. Dealing now with the key element, emotions themselves, there is an obvious effort in science to identify what the core emotions are that drive our behaviour. This is because, although we use various names to describe our internal states in our everyday lives, scientists discovered that there are actually only a small number that dominate our brains and bodies. Knowing those, and how they relate to each other, can help us deal more effectively with everyday personal and business challenges. Both Ekman and his teacher Tomkins studied facial expressions to reveal basic human emotions. Ekman's (2007) model of universal facial expressions found that, all around the world, six basic emotions are mentioned – anger, disgust, fear, happiness, sadness and surprise – while Tomkins' (2008) model, originally published in 1962 and 1963, identifies nine basic emotions or 'affects' as he calls them, some in low/high pairs, with each pair having a low/high expression, which scholar Nathanson calls 'hard-wired, preprogrammed, genetically transmitted mechanisms that exist in each of us' (Nathanson, 1992). These are:

1 Enjoyment (low), joy (high). This is a positive reaction to success, and the higher the emotion the stronger the willingness to share.

2 Interest (low), excitement (high). This is a positive reaction to a new situation and the higher the emotion the stronger the participation and engagement.

3 Surprise (low), startle (high). This is a neutral reaction to sudden change that can reset our impulses.

4 Anger (low), rage (high). This is a negative reaction to threat and the higher the emotion the stronger the physical and/or verbal attack.

5 Distress (low), anguish (high). This is a negative reaction to loss leading to mourning.

6 Fear (low), terror (high). This is a negative reaction to danger and the higher the emotion the higher the impulse to run or hide.

7 Shame (low), humiliation (high). This is a negative reaction to failure, creating the need to review one's own behaviour.

8 Disgust. This is a negative reaction to a bad offering, not necessarily restricted to food, and it motivates the impulse to expel and reject.

9 Dissmell. This is a negative reaction to a repelling situation and it strengthens the impulse to avoid and keep things/people at a distance.

Both Ekman's six and Tomkins' nine basic emotions have influenced both scientific and non-scientific professionals around the world, mainly because of their straightforwardness and direct relation to the human face. Those sets of basic emotions involve both macro and micro expressions of the face, the latter not able to be consciously detected or manipulated, and thus reveal deeper brain structures and how they impulsively react to stimuli. The latest research on Ekman's model reduces the number of basic emotions in facial expressions down to just four: happiness, sadness, fear/surprise, anger/disgust (Jack *et al*, 2014). Utilizing the latest technology and methods, this study demonstrated that happiness and sadness show unique facial signals over time, while fear and anger share initial signals with surprise and disgust respectively. Although, as the cycle of an expression matures, those pairs separate again and it is evident that facial signals are designed by both biological and social evolutionary pressures to optimize their function as expected by evolutionary predictions. So, fear and surprise start the same, as do anger and disgust, probably highlighting that when it comes to dealing with negative stimuli imminently compromising our survival, fast reaction and quick signalling is of elevated importance.

The basic emotions and the relations between them are intrinsically linked to chemicals in the brain. Hugo Lövheim's (2012) model is the first one to comprehensively link comparative levels of the three monoamine neuro-transmitting chemicals in our brain: serotonin, dopamine and noradrenaline. Lövheim combined Tomkins' nine basic emotions with all three neuro-transmitters and showed which chemicals have to be high and which low for every emotion. Serotonin for example, a chemical related to being able to control our behaviour, think clearly, regulate our mood and especially avoid aggression (Badawy, 2003), is high in the positive emotions of joy and excitement as well as in surprise and disgust. This is probably because in such emotional states we have the least need to show immediate aggression and the most to show openness and good decision-making. On the other hand, dopamine, the arousal hormone that gets us ready for action through anticipating results, is high in the two positive emotions of joy and excitement but also in the negative emotional pairs of fear/terror and anger/rage. This shows that dopamine is not just for positive (reward-related) situations but also for negative ones that need our immediate prediction of possible outcomes and appropriate response (Schultz, 2002). Finally, the stress hormone, noradrenaline (also known as norepinephrine), is high on distress, anger, interest and surprise, an interesting mix of negative, positive and neutral emotions. This is because noradrenaline is present at its highest when the situation we face is drastically different from the one our memory and expectations prepared us for, thus we go through 'unexpected uncertainty' (Yu and Dayan, 2005). So surprise, interest, anger and distress are naturally linked with high noradrenaline. In such cases we need to re-examine what we know and maybe learn new methods and behaviours. Lövheim's model helps us look deeper into the neurobiology of emotions and understand the brain changes on a chemical level occurring when we experience them.

Our premise is that, regardless of the model followed, leaders and managers need to understand the basic emotions and separate them from feelings, moods and styles in order to be able to work with them more effectively in various challenges. Being aware of the automatic reactions emotions have on behaviour, leaders can better steer their actions and the actions of others. The oversimplified and sometimes offensive model of the carrot and stick as a caricature representation of the avoid/approach systems in our brain has to be urgently replaced by a more realistic and complex knowledge of emotions and how they actually affect our motivation.

APPLYING TOMKINS' MODEL IN ORGANIZATIONS

We have been applying Tomkins' model of nine core emotions for some time now with our clients and the following key insights on each one are readily available for implementation by any manager and aspiring leader. We have re-modelled it to fit business and institutional environments.

Dealing with positive emotions

1 *Celebrate.* With any success, even the smallest, organize a celebration both for yourself and your team. Although the size and type of celebration should be directly related to the size and type of success, never allow a success, regardless how small, pass without celebrating it. Share the joy and your team will share it inside and outside the organization with multiple positive effects. Celebration is a social process, not an individual one.

2 *Explore.* In accordance with the growth mindset described in the previous chapter, new data, people, situations, etc, should be looked at with genuine interest and even with excitement. According to The Tomkins Institute (2014), such an approach directly connects effective thinking with a good emotion since this is the process by which learning is rewarding. Improve your thought and memory by excitedly exploring the world around you, constantly. This is the safest way to enhanced engagement. Furthermore, Jaak Panksepp, the world authority in affective neuroscience, has indicated that in his model of seven primal emotions – seeking, rage, fear, lust, care, panic/grief, and play – seeking is probably the strongest one since it is the seeking circuits that fire in the brain when we get excited by new intellectual connections, novel ideas and cutting-edge technologies, and when we enthusiastically search for meaning (Panksepp and Biven, 2012). That is, seeking keeps us motivated in the direction of what we seek for.

Dealing with the neutral emotion

3 *Stop-and-think.* When experiencing sudden, brief, unexpected changes it is advisable to refrain from a knee-jerk response and consider the situation closely before action. Especially when looking momentarily inside you fails to produce convincing insights quickly (based on memory and experience), stopping and giving yourself and your team time to think and better consider the situation is the best strategy. Importantly enough, acknowledging this fact to ourselves and the team makes us not weaker leaders but wiser ones.

Dealing with negative emotions

4 *Recharge.* Businesses and organizations of every type face constant changes and increasing threats from all around. Both geographically and industrially speaking, new competitors may arise from anywhere and threaten the very existence of even established corporations. Threats also exist within companies, with people trying to develop their careers at the expense of others. The brain reacts with anger when there are too many such threats and/or a few big ones. It stems from too much firing of neurons in the brain and the resulting inability to resolve the challenge at hand effectively. So, when feeling rage and wanting to attack in full force, just ask yourself if this is actually all about an overburdened brain and if attack is really the best way forward. Wanting to attack reveals that the one under attack is actually you, and so you need to reorganize yourself and the team to make sure your response will be optimized. It also shows that your preparation and/or attitude were not calibrated enough for what happened.

5 *Alert.* Similarly with above, when the situation seems unfavourable the brain can go into distress, with a natural impulse for mourning. The feeling that something is going wrong is a great opportunity for leaders and managers to reach out and alert their colleagues (and themselves, of course) that a different course of action is needed. Keeping quiet during distress or trying to ignore the problem are the worse alternatives because very little can improve like that, if anything at all. Alert, as with celebration, should always be a social response.

6 *Recompose.* Fear, or terror in the high end of this emotional state, is immensely beneficial when we face life or death situations since it focuses all our attention on the threat in front of us. Although fight or flight is the classic response to fear, accepting defeat and retreating fast are the most common ones suggested in Tomkins' model. Fear is a great tool for heightened attention, but only for very brief periods of time. This is because it swiftly depletes many other vital brain functions since it redirects all energy to attention and muscle tension to respond to a potential deadly threat. Many managers who like fear, as in the opening case of this chapter, probably do so because it allows them to laser-focus themselves and the people around them on a task at hand. This comes with significant side effects, though, and will backfire when not used extremely carefully. Fear releases the steroid hormone cortisol which, if triggered consistently, has devastating effects: killing brain cells, shutting down the immune system and disturbing sleeping cycles, among others (Brown *et al*, 2015). According to the authors of the book *The Fear-Free Organization: Vital insights from neuroscience to transform your business culture*, 'it's like driving a car with one foot on the accelerator and one on the brake' (Brown *et al*, 2015). Recomposing ourselves and re-establishing our strategy by realistically examining threats with the help of those closest to us can be of great assistance when facing fear.

7   *Reassure.* Shame in Tomkins' model is more about hiding a feeling than expressing one. It is more about someone substantially lowering your joy rather than creating a totally new emotion. Leaders can face failure by retreating into their own selves, closing communication channels and deeming the situation helpless. Or they can use this emotion as a serious signal to review their behaviour and their overall stance in order to take appropriate measures for improvement. It takes willpower and a strong growth mindset to do, but results can be spectacular. Reassuring yourself and your team of your collective abilities and plentiful opportunities for future success can work wonders for dealing with shame.

8   *Double-check.* Disgust is about having 'swallowed too much' of something that does not fit with our own values, attitudes and perspectives. It is an impulse reaction to reject things, people, ideas, situations we cannot digest, metaphorically speaking. Instead of showing contempt and alienating people, when feeling like instantly expelling someone or something it is far better to double-check where this response comes from. Is it justified or is it because of burnout-induced cynicism, stereotypes and other biases, and misunderstood information? The same goes for dissmell, which is literally about pulling ourselves away from something as rapidly as we can because of its potentially toxic effect on our well-being. The impulse is to create as much distance as possible, as fast as possible. This emotion can take over

when we momentarily consider a situation as 'toxic' and we want to automatically move back. Again, double-checking the reasons for this impulse is important to determine with a degree of certainty if this is the best action (which it could easily be) or if we are just fooled by circumstances and our own biases.

All core emotions need to be understood and embraced by modern leaders. No emotion should be ignored, suppressed or voided since emotions have evolved over millennia to ensure our survival, both physically and socially. As Tomkins suggested, though, our role is to increase the impact of the positive emotions over the negative ones. Utilizing all emotions, even the negative ones, for elevating our personal performance and the performance of our team is what separates inspirational leaders from managers that demotivate.

ACTION BOX: IDENTIFY THE EMOTION

Following the suggested model above, try to think of specific situations with intense emotions that you have experienced recently. Try to write down some basic keywords that your brain could use in order to manage the emotions that emerge based on the key insights above. From now on you can use these key words in order to activate and deal with the emotions in different situations.

The face of the leader

Without any doubt, one of the strongest emotions that influence the way that we behave is fear. We have evolved to fear certain things because we have realized biologically the consequences of particular actions. For example, if I had been bitten by a dog when I was young, there is a good chance I would develop a specific phobia about dogs, influencing my behaviour towards these (without any doubt) lovely pets. But now consider why some people are afraid of dogs. Have they all been bitten by dogs? Of course not. Then what? It is simply because we transmit fear emotions to others (Mallan *et al*, 2013). Fear and other emotions are transmitted through the social learning process. In other words, we learn from others to fear specific things and situations without having experienced any negative consequences of actual things and situations. If we see a role model (a parent, a friend, a teacher, a leader, etc) react with fear when they see or experience something, we tend to imitate this attitude (Burnett, 2016). Therefore, emotions are mainly transferred to each other through communication. Humans as social animals use different ways of communicating. However, non-verbal communication seems to be much stronger than verbal (Ten Brinke *et al*, 2012; Phutela, 2015). Facial expression is a very powerful way to communicate fear and other emotions. Evolutionarily speaking, this is very useful. Being able to 'read' fear on others' faces helps us to respond to danger, and this is extremely important.

This fact has enormous implications in leadership. Experimental cognitive psychology found evidence for evolved behavioural mechanisms including leadership (Vugt, 2018). There are studies confirming that there are some cues through which people recognize others as leaders just by looking at their faces. For example, a dominant-looking face of a leader could be enough to convince followers especially in a crisis to follow what he/she decides (Ohlsen *et al*, 2013). In addition, it seems that people show greater loyalty to a more masculine looking leader in hard times, while they prefer a feminine looking leader in more regular times (Spisak *et al*, 2012). The 'face' of the leader is so important to the perceptions of followers that it seem to influence the judgements of both adults and children. In a study by Todorov *et al* (2015), based on a limited exposure to faces, both children and adults were able to understand specific features of leaders. In a similar vein, another study by Olivola *et al* (2014) revealed that people can identify business, military and sports leaders just by observing faces. Interestingly, the same study argued that we cannot easily understand political leaders by observing their faces, which might be due to politicians' increased ability to manipulate our perceptions. Finally, emotions seem to be easily identified in the leader's face and influence our perception about him during action. For example, Trichas *et al* (2017) analysed videotaped presentations of leaders coordinating and concluding a team meeting, and found that the expression of happy emotions by leaders positively influenced the perceptions of the followers about their leaders.

At this point it is critical to mention that the ability to recognize and therefore influenced using emotions transmitted through the leader's face seems not to be moderated by the context, especially the cultural one. The truth is that in a real situation faces are rarely observed in isolation of other factors, like voices, postures, other faces and the general cultural context (Ekman, 1984). However, there are several studies confirming that the face of the leader is culturally independent, meaning that prototypes of leaders recognized though faces are similar across various cultures (Berggren *et al*, 2010). It seems that the face of the leader is global and well recognized. This implies that leaders should understand that their faces can be used as indicators and transmitters of emotions that lead to judgements like trustworthiness, dominance, etc. Therefore, the question is, can leaders develop the capability to transmit the right emotions through their faces? A study by Antonakis and colleagues (2011) gave a positive answer, supporting the idea that leaders can develop this capability after being trained in how to talk and use facial expressions and body language with particular audiences. Overall, understanding how emotions can be communicated by our facial expressions is essential for modern leaders.

Elementary, Dr Plutchik!

Understanding and dealing effectively with core emotions are the first necessary steps in becoming a better leader. The next, and more advanced, step is to understand how emotions are combined in order to create the wealth of feelings we experience in everyday life. In this direction, Dr Robert Plutchik, in the beginning of the 1980s, went beyond the identification of basic emotions, suggesting that each emotion, when combined with others, produces new ones and that each one forms the opposite of another emotion. By studying Dr Plutchik's emotional combinations we get a more complex but also a more representative view of the function of emotions in ourselves and others.

Starting with his view of the basic emotions, Dr Plutchik (2001) suggested that each of his eight core emotions has an opposite one: joy has sadness, trust has disgust, fear has anger, and surprise has anticipation. Knowing the opposite of an emotion is important, because the moment you experience a basic emotion you can tell which one is on the other side and take necessary actions. For example, when you do not trust a colleague enough and you want to improve the situation, you need to lower disgust, as that leads to rejection and expulsion. Double-checking the facts, re-examining your perception and identifying in yourself the reasons why this person's offering seems to challenge your own well-being are a good start. Sharing aspects of your findings in a constructive and positive way with your colleague and working together to improve trust between you is the recommended way forward. The same goes for surprise and anticipation. Fast-moving changes in all industries and the huge impact of complexity in our working lives has made anticipation a difficult emotion to sustain. Constant anticipation can lead to certainty, and certainty means security and control. Since these are in short supply in modern business environments, if we do not want ourselves and our people to be constantly surprised (an emotion that can easily develop into a negative one) then we should do our best to minimize creating rigid anticipation for specific outcomes as a precondition of well-being, personally and collectively (in the team). The small/purposive bets approach discussed in Chapter 3 can be very useful in avoiding big certainties and rigid anticipations. A strong sense of purpose, again from Chapter 3, can also help to develop healthy and open anticipation and thus mitigate constant unpleasant surprises.

But Dr Plutchik went even further, suggesting that putting two emotions together creates new ones, and that knowing these ingredients of an emotion can help us deal with this emotion better. The main combinations are:

- Joy and trust create love (A). The opposite is remorse (B), created by sadness and disgust.

- Trust and fear create submission (C). The opposite is contempt (D), created by disgust and anger.

- Fear and surprise create awe (E). The opposite is aggressiveness (F), created by anger and anticipation.

- Surprise and sadness create disapproval (G). The opposite is optimism (H), created by anticipation and joy.

A scary boss who brings results usually leads through submission and fear, very much like in the opening case of this chapter. In our experience, submission unfortunately is dominant as the leadership emotion of choice in many organizations globally. Regardless of the advantages of obedience and predictability from strictly followed rules, submissive behaviours can be extremely dangerous. It is not only that creativity is critically affected and that groupthink thrives in a submissive environment, but behaviours can easily take a darker turn, too. As shown in the seminal experiments of Stanley Milgram and Philip Zimbardo, obedience can lead to immoral actions since people suspend their own moral and rational judgement under a submissive authority. Milgram (1974) conducted his experiment in the early 1960s, where participants, acting as teachers, were asked to cause pain to other participants, acting as students, by punishing them with electric shocks when they made learning mistakes. The students were actors who never received electric shocks and the electric shock generator was faked, but the teacher-participants were ignorant of this. Many teachers (up to 65 per cent) chose to administer electric shocks even of very high intensity when the experimenter demanded they do so in order to follow the protocol of the test correctly. In Zimbardo's (2007) infamous Stanford Prison Experiment, conducted in 1971, participants were asked to live isolated in a prison environment and to act as guards or prisoners depending on the team they were assigned to. After just six days the experiment had to stop due to the extensive abuse that prisoners received from both guards and fellow prisoners on the guards' request. Those two classic experiments highlight that submission and obedience can easily transform some normal people into instruments of torture. It also demonstrates that the use of power/ authority is something that by default involves emotions that make people change their attitudes towards others. In organizations, this means that the use of a combination of fear and trust to lead people can get out of control and create an immoral environment in which cheating, lying and abusing becomes the norm. In a recent study, interviews with managers and employees from different companies operating in business environments that are under financial crisis were conducted (Psychogios *et al*, 2019). The results of this study showed that managers were more willing to use the above behaviours in order to manage employees' reaction to changes implemented due to critical situations. In other words, the nature of the context, like for example the CoVID-19 pandemic crisis in 2020, can enhance followers' submission and obedience to their leaders. As we have argued from the very beginning of this book, organizations need engaged, passionate, creative personnel, which submissive behaviour can never help us develop.

Love, interestingly enough, is the combination of trust and joy, since excitement accompanied by faith in someone creates a strong bond. Love and submission both have trust as a key ingredient, but joy replaces fear to create love. This is a key point in how Plutchik's (2001) analysis of emotions can be of great assistance for leaders. Trust is crucial for developing followership, but accompanying it with fear or with joy can result in very different kinds of followers! We all need to dissect emotions and see the best combinations that can be used to motivate both ourselves and others.

BASIC EMOTIONS AND PERSONALITY TESTING

More recently, there have been attempts to combine the neuroscientific view of emotions with more mainstream personality tests. Jaak Panksepp, the father of affective neuroscience, has developed his own system of basic emotions (Pankdepp and Biven, 2012), separating emotions into positive, which are lust, reward seeking, care and play, and negative, which are fear, rage/anger and panic/sadness. He also grouped them in terms of brain structure, with fear, lust, rage/anger and reward seeking being older emotions, deeper inside the brain, and care, panic/sadness and play being more recent, social emotions residing in middle brain. Together with Christian Montag, Panksepp reviewed the academic literature on personality testing and concluded that such tests need to be updated by incorporating robust scientific insights from the neuroscience of emotions in order to increase relevance and credibility (Montag and Paksepp, 2017). The famous Big Five personality test, which examines openness, conscientiousness, extraversion, agreeableness and neuroticism as the main traits, correlates with Panksepp's basic emotions in the following ways:

- openness correlates to high seeking;
- extraversion correlates to high play;
- agreeableness correlates to high care and low anger;
- neuroticism correlates to high fear, high sadness and high anger.

Conscientiousness does not have any correlates in basic emotions currently and this is to be determined by future research. In essence, using psychological models that do not consider in an integrative and pivotal way the neuroscience of emotions is an outdated and incomplete approach. Emotions, at least basic ones, should be considered above everything else.

ACTION BOX: DECODE PLUTCHIK'S EMOTIONAL COMBINATIONS

Defining and understanding emotions is not easy because, as Plutchik admitted, there are more than 90 definitions in relevant literature of what they actually mean. However, you can drastically improve managing your emotions by spending some time considering and applying his combinations. You can do this by taking each combination, from A to H, separately and trying to relate them to real-life examples from your own working environment. Try to see what will happen if you increase or decrease any of the basic ingredients. Most importantly, identify ways by which this is possible for you personally. For example, aggressiveness consists of anger and anticipation. If you feel constant aggressiveness towards a colleague, a department or a procedure, is it possible to lower its negative impact on you by lowering or changing your anger, your anticipation or both? How would any of this emotions strategy be possible?

In taking the concept of emotional combinations one step further, the entrepreneur, speaker and author Chip Conley (2012) suggests a number of emotional equations, as he calls them, that can help us achieve our professional and life goals more efficiently and effectively. Following Plutchik's rational of putting two emotions together to create something new, Conley created his own elaborate formulas. Three of them are congruent with our own viewpoint of modern, brain-based leadership.

1 *Curiosity = wonder + awe.* We have already discussed curiosity as a prerequisite for creativity as well as its importance in always asking questions, allowing the brain to process new information and challenge old biases. According to Conley, curiosity consists of two main ingredients, wonder and awe. While wonder is the pure excitement of discovery, awe is fear and surprise combined. In order for our leadership curiosity to work well, we need to allow the pleasure of being faced with something potentially amazing, a new product/service, colleague, technology, etc, to go hand-in-hand with humility and the feeling of being part of something bigger than ourselves. Awe allows us to connect with the world around us and adopt a more realistic view of the situation without losing the aspect of surprise when experiencing something grandiosely new. Curiosity without wonder is dry, shallow and short-lived, while curiosity without awe is full of arrogance, isolation and misinterpretation.

2 *Regret = disappointment + responsibility.* Within dynamic business environments leaders are expected to take many decisions daily. Although some of these decisions have higher weight than others, choices are usually plentiful for each decision, and choosing one alternative over others can emotionally backfire. This equation suggests that the higher the disappointment, your responsibility, or both, the higher the regret. Regret itself, according to Conley, is not necessarily a damaging emotion,

but if it turns to remorse (extreme regret) then it can be devastating. We have seen through our experience that regretting a decision can be beneficial as far as it leads to deep learning and corrective future behaviours. However, we have also seen regret becoming a norm, driving managers to question their skills and damaging their decision-making ability. Regret in leadership is unavoidable. Minimizing your disappointments by adopting a long-term life view, searching for lessons in failure, delegating responsibility whenever appropriate and being always aware of the big picture are effective strategies to address this equation.

3 *Thriving = frequency of positive/frequency of negative.* Thriving, or positivity as Conley also calls it, is about the relation between positive and negative events in our lives. Better put, it is about our perception of them. In order for this equation to have a beneficial outcome, positive events have to outnumber negative events by three to one. This is because negativity has a stronger pull on us than positivity, so positive events have to always be more than the negative for the equation to really work and to lead to thriving. Evolutionarily speaking, we have developed a higher sensitivity to negative stimuli than to positive, and this can be easily seen in the fact that in all models of core emotions the negative ones always outnumber the positive. In our primitive times, negative emotions ensured survival more than positive ones and this is why we tend to notice the negative more than the positive. Yet, in today's safer environments it is positivity that leads to healthier and more sustainable motivation and thus to desired corporate behaviours that can ensure creativity, engagement and growth. Changing our perspective to start noticing more positive information around us, making sure we operate under the right values, appreciate any lesson from any event, and connect ourselves to other positive people can help us increase the nominator and decrease the denominator in this equation.

All these equations, together with Plutchik's emotional combinations, can have a positive or negative final effect on your leadership capabilities. It is your decision which one it will be. You have enough material now to start working on your emotional toolbox to create your own mix of emotional ingredients that will help you achieve personal and organizational objectives. Nevertheless, there is a specific positive emotional state that has been ignored by organizations for far too long. Although we all try to maximize it in our private lives, it seemed, and still seems for many, inconsistent with management. This is happiness, and its advantages will surprise all those still believing that workplaces have to be charged with negative emotions such as fear for people to optimize their performance.

Bliss leadership

Claudia Hammond (2005), in her book *Emotional Rollercoaster: A journey through the science of feelings*, mentions that happiness is usually at the top of the list when people are asked to name emotions, and that it exists in almost all known models of core emotions (often referred to as joy by scientists in order to capture more clearly its momentary expression). Based on this observation, it is remarkable that psychology and neuroscience were preoccupied for so long with our mind's problems and not with its positive side. This is probably another proof of our brain's negativity bias, as mentioned earlier. It was only in 1998 that the then incoming president of the prominent American Psychological Association (APA) Martin Seligman announced that psychology had to focus more on the positive side of our mind and called for his term's focus to be positive psychology (Wallis, 2005). It was about time that scientists' obsession with negativity gave way to a more balanced approach that included all those elements that make our mind thrive, not only those that make us mentally ill.

I CAN'T GET NO... HAPPINESS

'I can't get no satisfaction' the Rolling Stones sang around the world in one of their most recognizable songs, first released in 1965. And they were right at that time. Management and marketing alike considered satisfaction to be a core measurement for employees and customers respectively. The idea was that if employees and customers are satisfied they will behave in the way they should; employees with productivity and customers with purchases. This managerial idea found its limit in one of the most popular management paradigms to emerge in 1980s – total quality management (TQM) (Psychogios, 2005). Satisfaction is mainly based on multiple discrepancies theories (Michalos, 1985) in which people's satisfaction depends on comparing themselves to multiple standards they hold, including other people, goals, ideal levels of satisfaction and past conditions. Satisfaction is higher when the comparison shows favourable results (downward comparison – set standards are lower than current personal state) and lower when comparison is unfavourable (upward comparison – set standards are higher than reality). The problem with this approach is that satisfaction alone cannot lead to the heightened behaviours we want ourselves and our employees to demonstrate. It seems that satisfaction has become, most of the time, a necessary but not a sufficient condition for increased engagement and boosted loyalty. In numerous studies we have done in our clients' organizations, using both quantitative and more exploratory qualitative methods, to help them identify key issues for performance improvement, it constantly came up that satisfaction was rarely related to any important factors. We believe that satisfaction has a new meaning today, mainly because of uncertainty. The latest global financial crisis, increased competition from around the world, fast technological advancements, corporate outsourcing and downsizing programmes, and even climate change create a landscape of the unknown that naturally makes many people feel unsafe and insecure. In such an environment, just having a stable job and getting paid regularly can, for many people around the world, be a good reason for being satisfied. It actually means 'I am OK at the moment, look around you', or 'I am OK at the moment, I do not have bigger ambitions', rather than 'We will do great!' Furthermore, and following Herzberg's renowned hygiene and motivation theory (Herzberg *et al*, 1959), we are observing that satisfaction itself has become a hygiene factor rather than a motivational one. This means that satisfaction as a measurable factor and explicit business terminology on employees' well-being is actually measuring basic acceptance at best and little more.

No great leader will be ever remembered because they created a satisfying working environment. Satisfaction is not enough. Great leaders are great because they can spread strong and heart-quickening positive emotions. So, it is Pharrell William's global hit song 'Happy', released in 2013, that is the soundtrack to modern leadership, and not the Rolling Stones' 'Satisfaction'.

To start taking seriously, and unlock the secrets of, positive emotions, it was necessary to move towards a more positive view of the mind. Happiness (or joy, or subjective well-being among other names) quickly rose to become the positive emotion of choice in both science and pop culture. This unfortunately means 'smilies' everywhere. But fortunately it also means significant research into the inner workings of happiness in our brains and behaviour. For both the wider society and organizations, the study of happiness brought a surprising and powerful insight. Happiness is not just an outcome of successful work, but a prerequisite. This counter-intuitive finding went against existing values since many cultures around the world perceive happiness as being the result of success. However, Boehm and Lyubomirsky (2008) have found substantial evidence in support of the reverse hypothesis, that is, happiness is the reason why some employees are more successful than others. By conducting a review of a large number of cross-sectional, longitudinal and experimental studies, they concurred that:

> Taken together, the evidence suggests that happiness is not only correlated with workplace success but that happiness often precedes measures of success and that induction of positive affect leads to improved workplace outcomes.

This means that inducing a positive atmosphere at work will not undermine success; it will actually cause it. Withholding positive feelings for only when success comes along will inevitably lower the chances that success will ever happen. And the evidence for this effect is widespread. Simon Achor, the global advocate for happiness at work, summarizes relevant research in his best-selling book *The Happiness Advantage: The seven principles of positive psychology that fuel success and performance at work* (2010) showing that:

- Optimistic salespeople did 56 per cent better at their work than pessimistic colleagues.
- Doctors in a positive mood showed three times more intelligence and were 19 per cent faster in diagnoses than others.
- Students primed (put in a mental state without knowing) to feel happy before taking exams did far better than those primed to feel neutral.

Such strong evidence, along with Achor's own empirical research on more than 1,600 high-achieving undergraduates at Harvard and his professional engagement with top companies around the world, led him to claim that:

> It turns out that our brains are literally hardwired to perform at their best not when they are negative or even neutral, but when they are positive. Yet in today's world, we ironically sacrifice for success only to lower our brain's success rate. Our hard-driving lives leave us feeling stressed, and we feel swamped by the mounting pressure to succeed at any cost.

And this cost, it seems, is success itself. There is a very interesting surprise here. In difficult times, when we personally and/or the company we work for are not doing very well, we allow the negative emotions to take hold of us and we postpone happiness, individually or collectively, for when things get better. This is exactly the happiness-averse mindset that makes success more difficult and more remote. What we need to do is the opposite. Regardless of the current conditions, we need to retain our high spirits and always strive for the best. This, mathematically, can lead to success more easily than negative or neutral emotions that we might find more fitting for difficult times. This is why we always try to create the best possible uplifting conditions when we work with a client and also to keep ourselves constantly upbeat. This is also supported by a recent study about organizations that are hit by financial crises. In particular, when investigating various case-studies of companies that struggle to survive in a crisis environment, through interviewing managers and observing behaviours and managers' and employees' interactions, it was found that where managers and leaders focused on the positive outcomes and emotions they achieved better results and the survival of their companies was more likely during the turbulent period (Psychogios and Szamosi, 2015).

Happiness sets the foundation of success in any project because of its ability to release intelligence, creativity, collaboration and commitment. We distinctively remember a situation when a team that was deemed ineffective sprang into frenzied action when a positive atmosphere was created after our recommendation and interaction with the team members. At the end of the year this initially forsaken team received the highest achievement award by the company's CEO. No person, no team, no company can categorically be called ineffective before a positive approach is wholeheartedly and thoroughly applied.

But how can we work towards creating more happiness at work for us and others in order to calibrate our brains for success? Professor and author Richard Wiseman, after reviewing the available literature on improving happiness, concluded that the fastest and surest route to a happiness boost includes the following three groups of actions (2009):

1 *Gratitude and appreciation.* Happiness starts from the inside out and thus it needs a sunny place to start from. Create this sunny place by periodically expressing gratitude for everything good that has happened to you. If possible, create a written account of them. Frequently refresh your memory about the fantastic experiences you've had and how they made you feel at that time. Try to recapture and relive the sensation within your mind and body. Examine your current situation and handpick only what is positive. Try a bit harder and you will be surprised by how many

positive moments go unnoticed because of the overwhelming and sticky power of negativity. Put aside the mind's negative veil and the sun will shine through. Finally, do not hold back on your appreciation for those close to you. Actively thank them for their contribution and efforts as well as for their personal support. When showing appreciation, try to be as specific and concrete as possible. This will benefit both you and the person you are showing appreciation to.

2   *Experience and sharing.* Our brains are hard wired to react to real life experiences better than to our material belongings. So, in order to boost your happiness fast, start engaging in refreshing and rewarding experiences as soon as possible. Also, create and offer such experiences, not just material rewards, to your team. The 2015 results of *Fortune*'s top 100 best companies to work for (Colvin, 2015) suggested that free perks and futuristic offices are not the decisive factor for a company to appear on the list but rather the ability 'to foster strong, rewarding relationships… among their employees'. So, it is the intangible nature of the amazing experience of working with the people you like that makes people happy in their workplace and not the tangible benefits. It is only natural, then, that sharing has been scientifically proven to be another strong booster of happiness. Regardless of most people believing the opposite, sharing an item or an experience makes us happier than when we keep things only for ourselves. As Professor Wiseman reports, even five non-financial acts of kindness per day can substantially increase our happiness. No surprise then that the concept of random acts of kindness have become so popular in management publications worldwide.

3   *Body language and behaviour.* Whenever we perform in class the famous happiness-inducing experiment, which includes holding a pen with your teeth, the results are astounding. If you hold a pen horizontally in your teeth without touching them with your lips for few seconds your brain immediately releases happy hormones just because it thinks you are smiling. Since you are smiling, something good must be happening! On the contrary, if you hold the end of the pen with your lips forming an O and not touching it with your teeth, your brain believes you are frowning and releases stress hormones. The amazing finding from this classic experiment is that happiness is sustained after the end of the test and makes people connect more positively and improves their memory of happy events. This, remarkably, shows that there is a two-way communication process between the brain and the body. It is not only 'what the brain

feels the body will show' but the other way around too. Smiling more, adopting a more positive, confident and upright body posture, using more positive wording when speaking and generally acting more happily will tell the brain to behave accordingly. Paraphrasing the adage 'be the change you want to see', we advise you to 'be the happiness you want to feel'.

THE 'THREE DS' OF HAPPINESS

In his book *Happiness by Design: Finding pleasure and purpose in everyday life* (2014), Professor of Behavioural Science Paul Dolan, outlined three key areas for improving our happiness levels: deciding happiness, designing happiness and doing happiness:

- *Deciding happiness* is about starting to notice more methodically the feedback of our decisions and our actions concerning our emotions. This feedback can have as a source both ourselves and others. The main problem here is that the speed of daily life, as most managers can easily confirm, does not always allow for reflection on how our different actions impact our inner self. We need to make time, though, and frequently ask both ourselves and others around us how our decisions affected us personally.

- *Designing happiness* is about creating the environment that will induce the right behaviour to increase our happiness. For example, if our New Year's resolution is to read more books, we need to leave more books around us in order to help our brain notice and pick up books more easily, set our internet home page to a book review site so that the default exposure we get is associated with our decision, commit to visiting a bookshop every weekend in order to have a concrete benchmark to follow, and join a book reading club in order to allow for social norms to help us achieve our goal.

- *Doing happiness* is about focusing more on experience than on material goods, spending more time in pleasurable and purposeful activities, dedicating more time to people with a constructive impact on our life and mood, and minimizing distractions such as constantly checking our emails or our social media notifications.

All three Ds are crucial for increasing our happiness and, as a consequence, the happiness of those around us.

Connecting the emotional dots

From an emotions-based point of view, if fear and negativity were motivation version 1.0, and happiness and positivity motivation version 2.0, there is a version 3.0 that calls for a holistic approach to utilizing emotions in our personal and professional lives. Regardless of the model, all core emotions have strong evolutionary reasons to exist. The mere fact we have them today proves that they were instrumental to get us where we are now as humans. We need, though, to be able to recognize them whenever possible in order to determine the benefits or dangers that derive from them. As Plutchik (2001) eloquently put it, 'emotions can sometimes fail in their adaptive tasks'. This means that not everything we feel is justified by real circumstances and not all we are inclined to do, pushed by our emotions, is beneficial for us and the people around us.

As the science of optimism, happiness and positive psychology came to improve the old simplistic models of behaviour such as 'the carrot and the stick' and unlock the huge benefits of having a positive outlook in life and work, more advanced approaches started emerging that would eventually improve the management of emotions. Todd Kashdan, a renowned expert on negative emotions, and Robert Biswas-Diener, a famous positive psychologist (2015), advocate 'wholeness' as the key to emotional success. As they argue:

> [I]t is high time that we reevaluated long-held beliefs of what is negative
> and what is positive, psychologically speaking. It is time for a new way of
> understanding of what it means to be mentally healthy and successful, to see
> both positive and negative as part of the larger, and more viable whole. This,
> then, is the Holy Grail of psychology, wholeness... Wholeness is to psychology
> what enlightenment is to spirituality.

They see signs of wholeness already appearing in corporate language when leaders talk about full engagement and optimal performance. Such language reveals that organizations now prefer emotional maturity and utilization of any emotion, rather than the older, simplistic and outdated categorizations of emotions as good/positive and bad/negative. For them, it is only wholeness that can lead to real emotional agility, which is not about avoiding negative emotions but about 'taking the negative out of them'. They cite Adler and Hershfield's (2012) study of 47 adults, which showed that those people experiencing happiness in psychotherapy at the same period as when experiencing sadness had better results on their well-being than those that were experiencing the one or the other emotion only. The study concluded that changes in mixed, or concurrent, emotional experiences are a prerequisite for overall emotional improvement. Simply put, one has to experience, and deal with, both positive *and* negative emotions rather than prioritizing the one over the other, as was the cases of motivational versions 1.0 and 2.0, which prioritized negative and positive emotions respectively. Such results are groundbreaking since they shatter the barriers between emotional categories long held in psychology, as explained earlier in this chapter. Yes, we do have emotions that can be generally called positive and negative, and neutral for that matter, but this should not lead to stereotyping them as good and bad. As we mentioned in the section 'Applying Tomkins' model in organizations' earlier in this chapter, each of the basic emotions has to be experienced and even expressed in appropriate ways in order to utilize them more effectively and optimize their benefits. Each emotion gives us important signals about the environment and our own acclimatization to it that we should never ignore and/or suppress. Leaders applying wholeness in emotions know what to keep and what not from every emotion and, above all, understand the value of all emotions to their motivation, flexibility and well-being.

THE POWER OF THE DARK SIDE OF THE FORCE

Negative emotions have received extremely bad publicity over the past decade. They have been blamed for almost all psychological conditions that decrease a person's ability to function normally in our society, perhaps not unfairly. What is unfair, though, is that the positive effects of negative emotions have been mostly ignored.

A notable example of this is pain. Marketing campaigns the world over promise us that our pains, physical and psychological, will disappear if we use this or that brand. Any inconvenience in life, regardless how small and trivial, any concern we have, the slightest fears and dissatisfactions, can be effectively addressed by buying a product or service. Although there is nothing wrong with this approach per se, it has a considerable effect on our perceived threshold in dealing with pain. But pain is not the same as harm. 'No pain, no gain' highlights this in the best way. Pain is almost always needed to escape our learning plateaus and comfort zones to improve and achieve more. Pain is also a good indicator that something is important and worth experiencing. It goes without saying that long-term or intense pain causes or indicates harm. In many cases, though, pain (physical, emotional, social or existential) should not be avoided, feared or just understood. It should be fully embraced and, most importantly, experienced. Frustration is invention, sadness is learning, stress is focus and fear is empowerment. In the right dose, of course, which makes it a very personal affair. Furthermore, 'do no harm' does not necessarily mean 'inflict no pain'. It means avoid causing long-term damage. Short-term pain can actually be extremely beneficial when tied to effort, sacrifice and dissatisfaction with the current situation. Anyone doing sports knows this very well already. So, why are we developing such a pain-averse society? As psychologist Brock Bastian (2018) highlights in his book *The Other Side of Happiness: Embracing a more fearless approach to living*:

> In order to find true happiness we need to embrace a more fearless approach to living, we need to approach our negative experiences on the front foot... we need to experience pain. By 'pain', I mean all the things that are not pleasure. I mean the anxiety of a significant challenge and the loneliness of failure. I mean the sadness of a relationship breakup, or the fear of our own mortality... They are necessary for happiness. Without pain, there is no way to achieve real happiness.

The same goes for stress, a usual culprit against corporate life well-being. In health, stress has become the root of all evils. However, stress is a necessary motivational signal in our brain and body that induces decisive action for improvement. Prolonged pathological stress is bad, of course. Stress itself, though, is a natural and normal condition responsible for change and personal growth. According to Kelly McGonical (2015), health psychologist and Stanford lecturer:

> [t]he latest science reveals that stress can make you smarter, stronger, and more successful. It helps you learn and grow. It can even inspire courage and compassion. The new science also shows that changing your mind about stress can make you healthier and happier. How you think about stress affects everything from cardiovascular health to your ability to find meaning in life. The best way to manage stress isn't to reduce or avoid it, but rather to rethink and even embrace it.

McGonical offers a variety of ways for turning stress into your best friend, including:

- finding deeper meaning in your life;

- applying clear humanistic values;

- helping others, regardless of how small this help is;

- considering the cost of avoiding stress;

- rethinking your stress responses as body preparation and excitation to tackle an important situation;

- separating challenges from actual threats;

- finding purpose in your actions that are higher than just self-interest;

- reaching out and connecting to wider groups and communities;

- adopting a growth mindset which sees the upside in adversity;

- communicating your successes in dealing with stress to yourself and to others.

Last but not least, a negative mood that transmits negative emotions to people in a team, a process described in Chapter 4 as emotional contagion, has been found to have positive effects in analytical thinking, regardless of its negative effect on creativity, effort and coordination (Barsade *et al*, 2018). This means that negativity can be used as a strategic tool to help the team focus appropriately on a specific situation, and for a limited period of time, and produce positive results that will benefit everyone involved. But definitely not in the systemic and permanent way in the opening case of this chapter.

Pain, stress, a negative mood and other similar conditions are allies, not foes, to anyone applying our brain-adaptive leadership approach. This is the way to embrace a more nuanced and scientific approach towards human motivation.

A deeper and more scientific view of emotions is necessary for leaders to perform better but it is also necessary for our society to improve as a whole. In this direction, even animated movies can help. *Inside Out*, created by Pixar Studios and released by Walt Disney Pictures in 2015, introduced the science of basic emotions to children and parents alike. In the movie, the basic emotions of joy, anger, sadness, disgust and fear are personified in the heads of the main characters. This personification works well in showing how emotions are the key sources of individual behaviour as well as in portraying the crucial and mutually dependent role of emotions in interpersonal relationships. In a world where the heart emoji, a highly emotional ideograph, was proclaimed the most-used English word of the year for 2014 in online usage by the Global Language Monitor (Gander, 2014), such wide popularization of science can have a very positive effect on our collective understanding and acceptance of the complexity and beauty of our emotional brain. Furthermore, the publicity surrounding the UN-supported annual Word Happiness Report, which ranks 156 countries in an index based on their happiness levels, named the World Happiness Index, the decision of the United Arab Emirates to appoint a Minister of State for Happiness and to launch the National Programme for Happiness and Positivity, as well as the wide popularity of Scandinavian and Japanese approaches to well-being, called *hygge* and *ikigai* respectively, reveal a major switch in both emotional education and policy. Growth, personal, corporate, national or otherwise, without long-term emotional welfare is not real growth.

Finally, we need to mention that emotions play a significant role not only for leaders but for followers as well. A study by Sanfey (2007) based on brain imaging reveals that when followers believe that they have received unfair treatment from their leaders this activates specific parts of their brain associated with emotions (anterior insula). This study suggests that the emotions of team members and employees play an important role in their decision to follow a leader. They also play a very important role in perceiving our leaders as good or bad (Vugt, 2018). So, if you think that your followers' emotions are not related to their opinion about you as a leader, think again.

BOOST YOUR BRAIN: IDENTIFYING MOTIVATIONAL VERSIONS IN THE WORKPLACE

Create a table and put the three motivational versions in three columns. In separate rows put all superiors, colleagues and employees that are positioned around you. Consider them individually and characterize each of them as high, low and medium for each of the motivational versions. The version that dominates a person should be scored as high, the one that appears from time to time as medium and the one that is almost or completely missing as low. Remember:

Version 1.0: Fear and negativity.

Version 2.0: Happiness and positivity.

Version 3.0: Wholeness and emotional agility.

While considering every person, try to remember specific instances that when accumulated indicate specific patterns of behaviour as a result of the preferred motivational version. At the end of the list, add your own name and do the same for yourself. Try to be as open as possible. Then ask at least two people you trust in the organization to rate you by explaining the concept to them. Request that they provide specific instances to support their judgement. What did you learn from the whole exercise? Do you or the people around need to change their version? In what way? And, how?

Keep in mind

Carrots, sticks, satisfaction and other single-emotion approaches are a thing of the past, incompatible with modern neuroscience. Brain chemistry and circuitry as well as the close study of physical expressions suggest that we have a number of core emotions that have evolved to perform significant tasks in order to increase our survival chances. Leaders and managers cannot perform their duties without understanding how these emotions work and for what reasons. First, you need to understand that your face plays a significant role in transmitting emotions to others. Second, familiarizing yourself with the number and nature of our basic emotions, unlocking the surprising relations between them, taking advantage of emotional equations, utilizing the power of positive psychology and finally putting together the entire puzzle of emotions with wholeness and emotional agility are essential for brain-adaptive leadership. Finally, both positive and negative emotions can help you to achieve leadership priorities. Above all, our brain is an emotional organ.

References

Achor, S (2010) *The Happiness Advantage: The seven principles of positive psychology that fuel success and performance at work*, Crown Publishing, New York

Adler, JM, and Hershfield, HE (2012) Mixed emotional experience is associated with and precedes improvements in psychological well-being, *PLOS ONE*, 7(4), https://journals.plos.org/plosone/article?id=10.1371/journal.pone.0035633 (archived at https://perma.cc/YU7C-5ES6)

Adolphs, R and Anderson, DJ (2018) *The Neuroscience of Emotion: A new synthesis*, Princeton University Press, New Jersey

Antonakis, J, Fenley, M and Liechti, S (2011) Can charisma be taught? Tests of two interventions, *Academy of Management Learning and Education*, **10**(3), pp 374–96

Badawy, AAB (2003) Alcohol and violence and the possible role of serotonin, *Criminal Behavior and Mental Health*, **13**(1), pp 31–44

Barsade, SG, Coutifaris, CG and Pillemer, J (2018) Emotional contagion in organizational life, *Research in Organizational Behavior*, **38**, pp 137–51

Bastian, B (2018) *The Other Side of Happiness: Embracing a more fearless approach to living*, Allen Lane, London

Berggren, N, Jordahl, H and Poutvaara, P (2010) The looks of a winner: Beauty and electoral success, *Journal of Public Economics*, **94**(1–2), pp 8–15

Boehm, JK and Lyubomirsky, S (2008) Does happiness promote career success?, *Journal of Career Assessment*, **16**(1), pp 101–16

Brown, P, Kingsley, J and Paterson, S (2015) *The Fear-Free Organization: Vital insights from neuroscience to transform your business culture*, Kogan Page Limited, London

Burnett, D (2016) *Idiot Brain: What your head is really up to*, WW Norton and Company, New York

Colvin, G (2015) Personal bests, *Fortune*, 15 March, pp 32–36

Conley, C (2012) *Emotional Equations: Simple truths for creating happiness + success*, Simon and Schuster, New York

Dolan, P (2014) *Happiness by Design: Finding pleasure and purpose in everyday life*, Penguin, London

Ekman, P (1984) Expression and the nature of emotion, *Approaches to Emotion*, **3**, pp 19–344

Ekman, P (2007) *Emotions Revealed: Recognizing faces and feelings to improve communication and emotional life*, Owl Books, New York

Gander, K (2014) Top words of 2014: The heart emoji named most used term of the year, www.independent.co.uk/news/weird-news/top-words-of-2014-the-heart-emoji-named-most-used-term-of-the-year-9948644.html (archived at https://perma.cc/2ASX-GTXV)

Greenfield, S (2015) *The Human Brain: A guided tour*, Weidenfeld and Nicholson, London

Hammond, C (2005) *Emotional Rollercoaster: A journey through the science of feelings*, HarperCollins, New York

Heiss, ED (2014) The MBA of the future needs a different toolbox, www.forbes.com/sites/darden/2014/10/01/the-mba-of-the-future-how-many-doing-what/ (archived at https://perma.cc/8VPL-7J55)

Herzberg, F, Mausner, B and Snyderman, B (1959) *The Motivation to Work*, 2nd edn, John Wiley, New York

Jack, RE, Garrod, OG and Schyns, PG (2014) Dynamic facial expressions of emotion transmit an evolving hierarchy of signals over time, *Current Biology*, **24**(2), pp 187–92

Kashdan, T and Biswas-Diener, R (2015) *The Power of Negative Emotion: How anger, guilt and self doubt are essential to success and fulfillment*, Oneworld Publications, London

Lövheim, H (2012) A new three-dimensional model for emotions and monoamine neurotransmitters, *Medical Hypotheses*, **78**(2), pp 341–48

Mallan, KM, Lipp, OV and Cochrane, B (2013) Slithering snakes, angry men and out-group members: What and whom are we evolved to fear? *Cognition and Emotion*, **27**(7), pp 1168–80

McGonigal, K (2015) *The Upside of Stress: Why stress is good for you, and how to get good at it*, Penguin Random House, London

Michalos, AC (1985) Multiple discrepancies theory (MDT), *Social Indicators Research*, **16**(4), pp 347–413

Milgram, S (1974) *Obedience to Authority: An experimental view*, HarperCollins, New York

Montag, C and Panksepp, J (2017) Primary emotional systems and personality: An evolutionary perspective, *Frontiers in Psychology*, **8**, p 464

Nathanson, DL (1992) *Shame and Pride: Affect, sex, and the birth of the self*, WW Norton, New York:

Ohlsen, G, van Zoest, W, and van Vugt, M (2013) Gender and facial dominance in gaze cuing: Emotional context matters in the eyes that we follow, *PLOS ONE*, **8**(4), e59471

Olivola, CY, Eubanks, DL and Lovelace, JB (2014) The many (distinctive) faces of leadership: Inferring leadership domain from facial appearance, *The Leadership Quarterly*, **25**(5), pp 817–34

Panksepp, J and Biven, L (2012) *The Archaeology of Mind: Neuroevolutionary origins of human emotion*, W W Norton and Company, New York

Phutela, D (2015) The importance of non-verbal communication, *IUP Journal of Soft Skills*, **9**(4), p 43

Plutchik, R (2001) The nature of emotions, *American Scientist*, **89**(4), pp 344–50

Psychogios, AG (2005) Towards a contingency approach to promising business management paradigms: The case of Total Quality Management, *Journal of Business and Society*, **18**(1/2), pp 120–34

Psychogios, A and Szamosi, TL (2015) Fight or fly? Rationalizing working conditions in a crisis context, 31st European Group of Organization Studies (EGOS) Colloquium, Athens, 1–4 July (Conference Proceedings)

Psychogios, A, Szamosi, L, Brewster, C and Prouska, R (2019) Varieties of crisis and working conditions: A comparative study between Greece and Serbia, *European Journal of Industrial Relations*, DOI: 101177/0959680119837101

Sanfey, AG (2007) Social decision-making: Insights from game theory and neuroscience, *Science*, **318**(5850), pp 598–602

Schultz, W (2002) Getting formal with dopamine and reward, *Neuron*, **36**(2), pp 241–63

Spisak, BR, Homan, AC, Grabo, A and Van Vugt, M (2012) Facing the situation: Testing a biosocial contingency model of leadership in intergroup relations using masculine and feminine faces, *The Leadership Quarterly*, **23**(2), pp 273–80

Ten Brinke, L, MacDonald, S, Porter, S and O'Connor, B (2012) Crocodile tears: Facial, verbal and body language behaviours associated with genuine and fabricated remorse, *Law and Human Behavior*, **36**(1), p 51

The Tomkins Institute (2014) Nine affects, present at birth, combine with life experience to form emotion and personality, www.tomkins.org/what-tomkins-said/introduction/nine-affects-present-at-birth-combine-to-form-emotion-mood-and-personality/ (archived at https://perma.cc/YV9R-ZGW2)

Todorov, A, Olivola, CY, Dotsch, R and Mende-Siedlecki, P (2015) Social attributions from faces: Determinants, consequences, accuracy, and functional significance, *Annual Review of Psychology*, **66**, pp 519–45

Tomkins, S (2008) *Affect Imagery Consciousness (Volumes I and II)*, Springer Publishing Company, New York

Trichas, S, Schyns, B, Lord, R and Hall, R (2017) 'Facing' leaders: Facial expression and leadership perception, *The Leadership Quarterly*, **28**(2), pp 317–33

Vugt, MV (2018) Evolutionary, biological, and neuroscience perspectives, in *The Nature of Leadership*, ed J Antonakis and DD Day, pp 189–217

Wallis, C (2005) The new science of happiness, *TIME Magazine*, **17**(1), http://content.time.com/time/magazine/article/0,9171,1015832,00.html (archived at https://perma.cc/HLQ2-GCY9)

Wiseman, R (2009) *59 Seconds: Think a little, change a lot*, Macmillan, New York

Wolpert, D (2011) The real reason for brains, www.youtube.com/watch?v=7s0CpRfyYp8 (archived at https://perma.cc/J6PC-7ZP4)

Yu, AJ and Dayan, P (2005) Uncertainty, neuromodulation, and attention, *Neuron*, **46**(4), pp 681–92

Zimbardo, P (2007) *The Lucifer Effect: Understanding how good people turn evil*, Random House, New York

SUMMARY OF PILLAR 2: FEELINGS

Leadership lessons

Continuously improve emotional style

Improve your self-perception of emotions by:

- describing the situation in detail;
- giving a detailed account of the behavioural response that the feeling produced;
- creating a list of events that had similar behavioural outcomes;
- looking inside yourself with minimum interruption and with genuine interest;
- listening to others carefully and trying to understand their viewpoint;
- constantly scanning internally and externally for emotional cues.

Improve your leadership style by transmitting emotions to others to improve their performance. Leaders that manage to spread their emotions better, especially positive ones, are perceived as more charismatic by people and thus their leadership effectiveness increases.

Develop both your internal and external self-awareness, which can improve your relationships with team members and increase your job satisfaction.

Be aware of moods

Keep your mood positive and with high energy.

Consider your mood often to make sure you give yourself the chance to experience the right emotion at the right time.

Be aware of the power of the basic emotions

There are nine basic emotions that leaders need to be aware of: enjoyment, interest, surprise, anger, distress, fear, shame, disgust and dissmell.

Deal with the core emotions through:

- celebrating any success for yourself and your team;
- exploring new data, people, situations and other possibilities;
- taking time to stop and think, especially when experiencing unexpected changes;

- pausing when you are angry to consider if this is the best way forward;

- alerting yourself and your colleagues when a different course of action is needed;

- recomposing yourselves and re-establishing your strategy by examining threats;

- reassuring yourself and your team of your collective abilities and opportunities;

- double-checking the reasons why you keep as much distance as possible from a situation or person.

Be aware of the influence of the leader's facial expressions

Leaders should understand that their faces can be used to indicate and transmit emotions, and that others (followers) will make judgements about their trust-worthiness, dominance, etc based on their facial expression.

Training on communication tactics can help you develop the face of a leader.

Develop emotional agility and happiness

Develop emotional agility by following the three groups of actions below:

- gratitude and appreciation;

- experience and sharing;

- body language and behaviour.

Happiness and positivity are directly connected with leadership capability. Improve happiness levels by using the 'three Ds' approach:

- deciding happiness;

- designing happiness;

- doing happiness.

Brain automations

06

Gut reaction, faster solution

I AM IN CONTROL, I WILL CHANGE

He is a revered and fearless negotiator. He always has been. His advanced skills for closing the right deal during the most difficult circumstances got him his current position as the CEO, his first time at the top position, in a mid-sized investment fund. The fund wanted someone that could quickly find the right pathway to growth, since the last financial crisis had driven it into problems that would not go away easily. Determined to succeed, he started applying everything he learned in his career from day one in dealing with both external and internal people. He renegotiated the already existing deals, identified new opportunities and quickly established valuable connections. He quickly made progress on the inside, too, turning around bad decisions of the past, cutting costs and streamlining operations. His early achievements were widely celebrated by shareholders, who trusted him almost completely. However, his quick wins soon turned sour.

His ability to negotiate by quickly finding the opponent's weaknesses and attacking with surgical precision served him well in his task-specific, highly specialized jobs of the past and in setting the early pace in his new one. However, this same ability started creating more problems than it solved. With the passing of time, it became evident that his shark-like approach to problem-solving backfired in developing and maintaining relationships. Many of his top people within the company resigned and changed jobs, and many of the remaining ones were thinking of doing so shortly. Although he seemed to perform very effectively and efficiently in the technical parts of the job, he somehow missed incorporating the human factor as effectively. When confronted by some of the shareholders on this he claimed the opposite: it was because of his deep understanding of human beings that he was so good in negotiations and in getting what he wanted. He knew exactly how to discover soft spots, like fears and desires, and how to press them in order to achieve the outcome that he wanted. He acknowledged the significant role of emotions in the whole process, both in him and in others, and this was why he could get the numbers he was striving for. He was not cold-blooded or cool-headed: he just thrived on manipulation.

One morning, his closest ally in the company, and his own recruitment choice a year ago for deputy CEO, resigned. The CEO, shocked from the news, called his deputy for urgent discussions. During a heated two-hour conversation the ex-deputy CEO told him that he could no longer work alongside him, because he simply could not tolerate the premeditated manipulation the CEO exerted on himself and others. 'People should be treated with respect and not with a different approach each moment in order to get what you want from them,' he yelled. Changing faces, emotions and words so easily, day in and day out, to achieve results is not something that is valued in leaders, he continued. The CEO, agreeing with these observations, pledged to change. He would do his best to treat people differently, be more consistent in his attitude towards others and put values over short-term gains. He was grateful that someone had told him so and was determined to be a better person, a better CEO, a better leader. He considered himself a very mentally strong person, in full control of his behaviour, and this was what he was going to do. He convinced his deputy to stay.

Two months later, however, the deputy has gone for good. The CEO never really changed, regardless of his commitment and even some initial sincere efforts. The irresistible pull of his deeply rooted behaviour proved much stronger than his understanding of the need to change. A year after his deputy left, his initial great results turned to losses and the shareholders very seriously consider the option of seeking a new CEO.

This case shows that the concept of consciousness in human thinking. Its role in what actually makes us human is central to neuroscience and psychology. Do we take conscious decisions fully aware of risks and benefits? Does our mind affect our behaviour directly and exclusively? Can we change whenever we want to? Do we make our own fate? Are we in control? If yes, then we only need to train our conscious minds through formal and informal learning processes, as we have been doing for centuries. But if not, then leaders need to look deeper to find effective ways to understand and influence better their decisions and actions. Because, as we now know, we are not in control of ourselves, at least at the extent that we believe we are. Our brain is.

The mind-controlling brain

The unconscious is the part of the brain function that contains feelings, in this context more accurately called emotions, and thoughts, more accurately called intentions, that we are not aware of, and it influences the way we behave (*Cambridge Advanced Learner's Dictionary and Thesaurus*). The power of the unconscious has long been emphasized by psychology, with Sigmund Freud pioneering our understanding of how our deepest and most unknown desires affect our everyday lives. However, most recent research in the field reveals that the unconscious is not the dark place that Freud envisioned. On the contrary, it brings about unique advantages. Its processing capacities form an integral part of our healthy, dynamic and interactive state of being. McGowan (2014) highlighted that 'the nature of the unconscious thought that emerges from contemporary experiments is radically different from what Freud posited', and that viewing the mind with its unconscious influences 'honors the unique experience of individual human beings – something often overlooked by the current medical approach to the mind'. Our unconscious seems to be our biggest strength as a species and not our darkest weakness.

The impact of the unconscious on our behaviour and decision-making is profound. Much more than we are willing to admit. As John Bargh (2017), the leading researcher of the unconscious mind, reported, his brother-in-law had this to say when Bargh explained to him the drastic influence of the unconscious: 'That just can't be John. I can't remember one time when I was influenced unconsciously!' Not being able to consciously detect, fully experience and directly control unconscious thinking processes makes the unconscious difficult to believe in: 'seeing is believing' goes the old proverb, suggesting that something not seen is something that probably does not exist. The arrogance of the conscious mind, or the 'self illusion' as prominent psychologist Bruce Hood (2013) calls it, is manifested by its own limited but strong perception that control and reality are exclusively within its territory. That could not be further from the truth. This is one of the most exciting and also challenging moments in our own work with strong-minded people within companies. How can people accept the fact that most of their life has been directed by the unconscious, when they really feel that they have taken their key decisions in full awareness and responsibility? But it is exactly that, a feeling, not a scientifically driven insight. Don't be fooled by your mind: it masks the, mostly, wonderful and intelligent processes that make you who you are.

The unconscious is always there, and by ignoring it we can only harm our leadership potential. In the most famous experiment on the primacy of the unconscious on decision-making, Benjamin Libet and his team asked participants to make simple decisions, like pressing a button or moving their fingers whenever they felt like it, within a timeframe (Libet *et al*, 1983). They also asked them to note their decisions at the time they were made during the experiment. By monitoring participants' brain activity with an electroencephalograph (EEG) and their voluntary motor activity with an electromyography (EMG), Libet and his associates revealed that the electric build-up in their brains for the upcoming decision appeared approximately a quarter of a second earlier than the moment when participants became conscious of that decision, and almost half a second before they acted. This *readiness potential* in our brains comes before we are aware of our own decisions. And if you think this is shocking, Soon *et al* (2008) increased this gap to 10 seconds for simple decisions, claiming that this delay 'reflects the operation of a network of high-level control areas that begin to prepare an upcoming decision long before it enters awareness'. More recently, a study by Koenig-Robert and Pearson (2019) found an 11 second gap between unconscious decision-making and conscious response. As Pearson stated (Goldhill, 2019):

> We believe that when we are faced with the choice between two or more options
> of what to think about, non-conscious traces of the thoughts are there already,
> a bit like unconscious hallucinations. As the decision of what to think about is
> made, executive areas of the brain choose the thought-trace which is stronger.
> In other words, if any pre-existing brain activity matches one of your choices,
> then your brain will be more likely to pick that option as it gets boosted by the
> pre-existing brain activity.

This means that our awareness is not there, primarily, to take decisions. It is there to receive the decision, as a strong signal, from the unconscious and to make it happen. This is actually the main difference between the conscious and the unconscious: the unconscious provides the deep-rooted, complexly calculated, emotionally based and behaviour-driving motivation, or intention, and the conscious provides the best way of implementing it successfully.

Although the original study by Libet generated a lot of controversy, with later studies claiming they had different results by repeating the experiment (Gholipour 2019), the mounting evidence directly points to the priority of the unconscious in decision-making and of the conscious in implementation. This evidence provides extremely detailed timings when this happens in the brain together with the topography of where this happens. For example, Fried and associates (2011) conducted a now famous fMRI study to discover very specifically and accurately the brain areas involved in taking a decision about movement and in initiating this movement prior to people being aware of it. They found that:

> a population of [just] 256 SMA neurons is sufficient to predict in single trials the impending decision to move with accuracy greater than 80 per cent already 700 ms [milliseconds] prior to subjects' awareness. Furthermore, we predict, with a precision of a few hundred ms, the actual time point of this voluntary decision to move.

The incredibly small number of neurons needed to predict action with an 80 per cent plus accuracy is telling of both how neuroscience is becoming better in zooming into the brain to uncover hidden processes and how unconscious processes lead to conscious behaviours with high precision. If you feel like you are living in an episode of *Black Mirror*, the popular British science fiction TV series, you are not alone!

WHAT, THEN, IS CONSCIOUS THINKING REALLY ABOUT?

To implement decisions made by the unconscious, the conscious mind is responsible for two main functions: planning and inhibition. Let's start with planning. Since the decision is already taken and the motivation to act is already there, the mind scans the terrain to identify objects and actors, and places them in relative position to us and to each other, in order to identify the best route to implement the decision that the brain has already taken (Kosslyn and Miller, 2013). It does this by using two distinct neural pathways. First, identifying the nature of elements around us uses a lower brain route, which includes our visual cortex, our temporal lobe and the lower part of our prefrontal cortex. Second, positioning those elements in their exact location uses a higher brain route starting, again, in our visual cortex but then going through our parietal lobe, which sits on the very top of our brain, and the higher part of our prefrontal cortex. Our mind's ability to mentally time-travel, in the past and in the future, also supports this planning process of 'who/what is there' and 'where exactly are they' in order to plan the best way forward. This is the navigability function of the conscious mind as opposed to the decision-making, goal-setting and motivation-creating function of the subconscious mind (Baumeister and Bargh, 2014). But planning would be impossible if our body reacted immediately and impulsively, implementing unconscious decisions directly and without the evolutionary benefits of navigation/planning. So, the second important function of the conscious mind, inhibition, is about putting the brakes on to stop premature action and to leave time to plan the best course of implementing the decision. The conscious mind, residing mainly in the neocortex of our brain, does not initiate action but has a kind of 'veto vote' against the deeper brain's decisions to act. It does not motivate but it stops motivation. It does not decide but it stops decisions. According to renowned neuroscientist Joseph Ledoux, it does not have a 'will' as in 'I have free will to do what I want', but it has a 'won't' as in 'I won't do this' (McGilchrist, 2012).

Planning and inhibition, followed by abstract thinking and awareness respectively, are the key processes of the conscious mind, and they can explain the problem of motivation and engagement in modern companies. The percentage of employees that were engaged worldwide in 2018 was an abysmal 15 per cent, with 67 per cent of employees being disengaged and 18 per cent being actively disengaged (HR Exchange, 2018). Although total engagement increased slightly from two years ago it still represents just a fraction of the workforce. Actually, there are more actively disengaged employees, those that harm the company intentionally, than there are engaged employees! This can be attributed to the wide misunderstanding of thinking and motivation. For the brain, decision-making and motivation to act are the same thing and originate in the unconscious. The brain quickly analyses a situation and decides to activate its approach or avoidance motivational networks to initiate a corresponding behaviour. This is real decision-making. The other type of decision-making that appears in our conscious mind is not related to motivation and behaviour, and this is why it does not help people engage. It is abstract, conceptual and theoretical decision-making that is not related with action. Companies' obsession with over-analysis, tiring justifications and endless numbers does a disservice to our brain's ability to engage. Managers wrongly deploy conscious thinking techniques such as factual-only communication, relentless argumentation and perpetual reporting to convince and motivate. They are targeting the wrong decision-making centre. If leaders are to motivate and engage, they need to connect with and convince the unconscious that will, in return, provide the desired real decision and the inevitable action. This is why, while everyone in a team seems to agree and fully understand the new managerial plan, few or none give their best to achieve it. In such situations, conscious minds were convinced but unconscious minds were not.

Our brain prepares to take a decision and nudge us into specific actions much earlier than when we become aware of it. And it does this for very good reasons. The unconscious helps us save energy (as we have seen in Chapters 1 and 2), react very quickly in life-threatening situations, and form attitudes fast about others based on previous experiences. Our brain's unconscious processes underlie the whole way we deliberate and plan our lives (Bargh, 2014). In his bestselling book *Blink: The power of thinking without thinking*, Malcolm Gladwell (2005) compares our *adaptive unconscious*, as it is known, to a giant computer that quickly and quietly processes the large amounts of data we need in order to keep functioning normally in our everyday lives. If we processed all this stimuli consciously we would be barely able to take any action at all. Gladwell quotes the renowned psychology professor Timothy Wilson, who coined the term adaptive unconscious, saying:

> The mind operates most efficiently by relegating a good deal of high-level, sophisticated thinking to the unconscious, just as a modern jetliner is able to fly on automatic pilot with little or no input from the human, 'conscious' pilot. The adaptive unconscious does an excellent job of sizing up the world, warning people of danger, setting goals, and initiating action in a sophisticated and efficient manner.

In support of this notion, psychologist and behavioural expert Gerd Gigerenzer (2007) has also called for a wider acceptance of the *intelligence of the unconsciousness*, which is fundamentally based on simple rules of thumb, which in turn are based on evolved capacities of our brain. These evolved capacities help us make snap decisions or fast judgements and to act accordingly, by providing signals in the form of gut feelings. Taking into account that our advanced language ability is positioned in the top part of our brain, the cerebral cortex, deeper and more ancient brain systems try to communicate with us through gut feelings, impulses, instincts and intuitions, since they simply cannot speak. And this deeper, adaptive, intelligent unconscious has many times been proven to be able to make better decisions in situations as demanding as choosing stocks to invest in and buying a new property (for more, see Wilson, 2002; Gladwell, 2005; and Gigerenzer, 2007).

This is not to say that unconscious thinking is always superior to conscious thinking and that we have to abandon analytics altogether. The scientific debate is ongoing, with both sides conducting experiments to show that trusting your unconscious mind is either more beneficial or more dangerous in complex decision-making, where there are a lot of factors determining the outcome and, even worse, most of them cannot be easily understood or cannot be understood at all. On the one side, Ap Dijksterhuis, Loran Nordgren and associates, supporting their unconscious thought theory, claim that unconscious thinking, or deliberation-without-attention, can be far more beneficial than conscious thinking when a problem is very complicated and multifaceted. In such situations, too much thinking leads to bias activation, analysis-paralysis, worse decisions and considerably less post-decision satisfaction (Bos *et al*, 2006). On the other side, scholars such as Ben Newell (Newell *et al*, 2009) claim the opposite, saying that only extensive rational thinking can bring results within complexity while fast, unconscious or unfocused thinking can lead to serious mistakes. Regardless of where you personally stand in this debate, the important point is that decisions do not just form from thin air. The brain spends considerable energy before a decision is made in order to prepare the conscious mind for it. The quality of this preparation will influence directly the quality of the decision, conscious or unconscious.

We usually do a very simple experiment when we want to demonstrate to our audiences the power of the unconscious will over the conscious one. We ask them to try *not* to think of a white polar bear for the next five minutes, which is almost impossible for most of us. Sooner or later a polar bear appears in our mind's eye. In fact, although we try to avoid thinking of a white polar bear by filling our conscious mind with other, different thoughts, our unconscious mind remains alert for any signs of the unwanted thought, the better to help us chase it away. This popular thought exercise, first mentioned by Fyodor Dostoyevsky in 1863 and proved in experiments by Daniel Wegner of Harvard University (Winerman, 2011), helps us to highlight the fact that our conscious mind is not as in control as we thought – if it is at all. Our unconscious is. The good news is that we can study the way the unconscious mind works and recalibrate it for the benefit of further developing our managerial and leadership skills.

ACTION BOX: EXPERIMENT WITH UNCONSCIOUS AND CONSCIOUS THINKING

Select an organizational problem of high complexity (involving a lot of factors either known or unknown) that your company would like to solve in the near future. Take your team and separate them into three smaller groups. Ask group A to take a look at the problem and make an immediate decision. Spending time in thinking of any kind is not allowed. Ask group B to do the opposite. They can take the data, add some more if needed, spend half an hour at least in advanced deliberations and then make up their minds. Ask group C to consider the problem briefly, but then distract them with any trivial game of your choice (solving anagrams, puzzles, angry birds, etc). They have to make their decision about the challenge after they finish playing the game and without further deliberation. You can even add a group D that has to sleep on the problem overnight and suggest a solution in the morning, without having any prior interaction or deliberation on the topic.

Observe any differences in the decisions made. Which decisions you would follow and why? Is it the time spent on rational deliberation that determines the quality of the outcome, or something else?

Reshuffle the groups and repeat the experiment with a problem of very low complexity (a more simple issue with fewer aspects and attributes). Do you see any differences between the two cases?

In the first case (complex problem) rational thinking and over-analysing all parameters of the problem may result in brain paralysis since the nature of the problem (complex) involved many unseen factors that cannot be easily detected by the conscious mind and therefore we need more intuition and activation of the unconscious mind. In the second case (simple problem) the conscious mind can easily take control since factors can be seen, explored and determined accordingly.

It's prime time

Automatic brain responses to real world situations are the norm rather than the exception in our everyday lives. Taking into account that 98 per cent or more of our total brain activity daily is completely unconscious (Gazzaniga, 1998) and that 95 per cent of our decisions are unconscious (Zaltman, 2003), this should not be a surprise. What is surprising, though, is the fact that many managers and leaders go about their careers being either oblivious or conceptually against the idea that their behaviour, and that of their colleagues, is highly driven by the unconscious. They need to understand this is true, and that this is not necessarily a bad thing.

We mentioned above the phenomenon of the brain's *readiness potential* in the process of taking a decision. Before we consciously make up our minds and move to a behaviour, our brain's neurons start firing within the unconscious realm, creating the basis of this decision to be made. This neural preparation process for decision-making is not independent of the external environment, and it can be highly influenced by stimuli. Influencing the readiness potential of our brain means influencing our decision-making and behaviour, so that we can direct or nudge our actions in a preferred path. The way this is achieved is called *priming*. Priming is an effect of our implicit memory in which exposure to one stimulus affects the response to another one (Meyer and Schvaneveldt, 1971). Priming is preparing our brain for specific thoughts and behaviours (Alter, 2013). The brain will unconsciously and very quickly consider cues that are detected in a situation to prepare our mind to take a suitable decision and to adopt a fitting behaviour. Different external cues will lead to different neural priming/preparation and possibly to different decisions and behaviour. In essence, messages of the environment can put our brains in specific thinking and acting pathways. In order to optimize our own performance and our team's, we need to make sure that we use priming in a positive and productive way. Otherwise, priming can work heavily against us, without us ever knowing it. The list of experimental results indicating the power of priming is long (McGilchrist, 2012):

- Participants in a general knowledge quiz were separated into three groups, each one engaging in different activities that would prime their brains into different mental states: one as professors, one as secretaries and one as hooligans. The professor-primed group significantly outperformed the other two in the quiz, with hooligans doing worse of all.

- Participants primed with a mental state of elderly people (reminded that the elderly are sentimental, play bingo and have grey hair, among other characteristics) became more conservative in their opinions, while others that were primed with a politician state of mind (reminded of the main characteristics of politicians) expressed themselves in a lengthier manner.

- When elderly people were primed with the positive association of being old, they performed much better in memory tests than those primed with the negative aspects of being old.

And that's not all. In a much publicized study, US voters participating in a priming experiment on American identity and willingness to vote at San Diego State University in 2007 considered the then US presidential candidate Barack Obama as less American than even Tony Blair, the former British Prime Minister! They were initially primed to consider Obama's ethnicity, labelled in the experiment as 'black' (Kristoff, 2008). Such negative racial biases can sway people's opinions in subconscious ways and thus need to be confronted openly in order to lose their undetected power over our decision-making.

In a series of three famed experiments, Bargh and associates (1996) found:

- Students primed by sentences with associations to the elderly population walked slower in the corridor after they were primed than those of a control group that read non-associative sentences.

- Students primed by sentences with associations to polite and respectful behaviour waited more patiently and for a longer time outside the office of a busy researcher (who was part of the experiment) than those primed with words such as 'rude', 'disturb' and 'aggressiveness'. Actually, 82 per cent of those primed for politeness did not interrupt the busy researcher at all.

- Among students primed by words that reminded them of their race, African American students reacted with more hostility to a researcher's frustrating requests. Again, as in the case of the Obama study, negatives bias in issues of race (even 'self biases' observed in this particular study) can lose much of their subconscious influence when people become aware of them.

All these results, and they are all worth noting, demonstrate that our brain automatically responds to external cues by subliminally adopting a specific frame or state of mind. Priming prepares the brain to take the 'right' decision and take the most 'appropriate' behaviour by evoking attitudes and models pre-established into our long-term memory.

Priming works best when people are not aware of its effect. When we become aware of the effect, priming stops working. This means mainly two things for leaders. First, when we detect in ourselves behaviours or reflex reactions that we do not like, we should seriously consider if they are the product of unfavourable priming to our brain. Are there any external cues that are automatically putting us in an unproductive and ineffective mental state? Second, we should do our utmost to participate in and/or create a working environment that is priming our brain with the best attributes for high performance and the highest chances of success. Are the companies we work for, our colleagues, even our offices boosting our brains for maximum performance or do they do the opposite? The core takeaway, though, is that our brain is not independent of its environment. It will continuously receive and process subliminally much more information than our conscious mind could ever handle and it will reflexively respond by altering our conscious mind's ability to make decisions. Neural readiness potential and mental preparation for decision-making through priming are what our brains have evolved to do very well. Leaders ignoring or opposing those scientific insights are missing the amazing opportunities of recalibrating their brains for greatness. This is because the knowledge of the unconscious mind's influence on our behaviour is the single most important weapon we have to shape it favourably for us. Knowledge is true power, in this respect. As Bargh (1994), a prominent advocate of our unconscious mind whose research and words are included prominently in this chapter, stated categorically on the matter:

> Automated social cognitive processes categorize, evaluate and impute the meanings of behavior and other social information, and this input is then ready for use by conscious and controlled judgment... [but] the unintentional and uncontrolled nature of automatic analyses of the environment does not mean they are impossible to control or adjust for when one is aware of them, if one desires.

Our unconscious mind is more powerful than our conscious one, in the sense that it consumes more brain power and analyses more information than our controlled thinking process. The challenge we all face in our jobs is not to find ways to stop this power, a very dangerous and neurologically almost impossible thing to do, but to harness it to our advantage.

Priming, like other influential concepts in science, has not been without opponents. Some experimental results failed to replicate in later studies (see Doyen *et al*, 2012 and Bargh, 2013), bringing priming under fire in an outspoken debate. In particular, these studies claimed that their experiments failed to show priming, suggesting that both priming and experimenters' expectations are instrumental in explaining social behaviour. The latter arguments triggered the renowned psychologist Daniel Kahneman (2012) to issue a public letter to fellow researchers in support of more research in priming, a concept he claimed he believes in. Even Doyen and associates, who were among those that did not manage to replicate priming results in the exact same way as in one of the famous studies, explicitly concluded that 'unconscious behavioral priming is real' (Doyen *et al*, 2012). Since then, the concept of priming has attracted even more criticism from academia, leading many psychologists to denounce it altogether (Chivers, 2019). However, the original advocates of priming, business practitioners and neuroscientists alike, continue to support the concept. New evidence suggests that priming does affect both the brain and our behavior, but in a way that is more consistent with a social group's pre-existing attitudes (Chivers, 2019), or at least, when it is not against them. This means that priming might not be the strongest tool for inducing complete and revolutionary behavioural change but could be effective in nudging smaller behavioral modifications that can, accumulatively and over time, lead to desired results.

There are plenty of leadership implications associated with the concept of priming. Kark and Shamir (2002) support the view that transformational leaders can possibly influence the relational self of followers by having specific behaviours that prime followers' self-concepts and identity. They claim that specific leaders' behaviours, such as psychological support and compassion, can prime a relational identity of followers and make them more connected to each other. In other words, the stronger connections and emotional bonds that leaders develop with their followers, the more likely it is that followers will be more relational in their attitudes towards others. Moreover, priming seems to work for developing someone's leadership potential. In an experimental study, Latu *et al* (2013) asked 149 students from both genders to give a public speech in a virtual reality environment. Participants were exposed to pictures of Hillary Clinton, Angela Merkel, Bill Clinton, or no picture. Female participants were exposed to pictures of popular leaders from both genders. They found that female participants tended to talk less when exposed to Bill Clinton's photo or no photo at all. But another group of female research participants, when exposed to pictures of female leaders (Hillary and Angela), showed significant increase in terms of the length and quality of their speech. In other words, female role models primed female research participants to act in a more inspired and meaningful manner. In short, managers' leadership behaviour can unconsciously influence the behaviour of their employees. But this is already known, for example when we say 'Lead by example'. Now we have strong scientific evidence for it.

All in all, automated cognitive processes, priming and readiness potential are concepts that are here to stay since our unconscious mind seems to be in higher – and more beneficial – control of our thoughts and behaviours than we ever expected. Having said that, we are not simple zombies in the service of our uncontrolled brain. Through awareness and knowledge we can reduce the negative effects of priming, such as stereotyping, and increase its positive ones in order to maximize our leadership and managerial performance. Nevertheless, those that still oppose the presence, power and often intelligence of our unconscious mind, hailing the unconditional superiority of our rationality, will be living in an illusion that is both ineffective and hazardous for all. Like in the opening case, we cannot simply go on ignoring the limits of our ability to be fully in control of our own behaviour.

PRIMING IN ORGANIZATIONS

We often find people in corporations and institutions that are blind to the interdependence between the brain, its environment and the resulting decisions and behaviours. Somehow, many are led to believe that performance and mental agility should be something isolated from external conditions, streaming solely from inside one's head and heart. As this viewpoint goes, it would be great if the company performs fantastically regardless of everything going on around one individual. We find this approach damaging because, above all, it removes the organization's and leadership's responsibility for creating the right environment for achievement. For us, the individual is always important as far as both the individual and those in the immediate environment are aware of and using effectively the key determinants of personal and social behaviour. As the psychologist Philip Zimbardo (2008), who conducted the famous Stanford Prison

Experiment, mentioned in the previous chapter, often explains, it is not the individual that should be always accused of a behaviour ('the bad apple' approach), nor even the immediate environment ('the barrel of apples' approach), but the ones that have the ability to design the environmental conditions where actors behave and actions take place ('the barrel maker' approach). True leaders are 'barrel makers', creating positive and dynamic environments that prime their own and their employees' behaviour for strong and lasting performance. Below are proven ways of creating the right priming conditions:

- *Culture.* Corporate culture, or even departmental or team culture, are strong primers for decision-making and behaviour. Cultural constructs are found to be key drivers of social behaviour since, when triggering the implicit knowledge of its values, people are automatically put in a specific frame of mind and make behavioural choices accordingly (Hong *et al*, 2000). This means that organizational culture with its actual working values and as an everyday-lived, collective experience – not as a polished presentation or online entry on the corporate website – determines subliminally the way that people's brains are readied for action. It is imperative that you unveil the actual work ethos in your organization and understand its impact on priming people for specific behaviours in every situation, like in meetings, negotiations, presentations, etc.

- *Context of message.* Messages are not independent of the context or media they appear in. Actually, our brains are primed to accept, reject or generally decode and assign meaning to a message based on the implicit perception of the medium, for example a person, an electronic newsletter or Facebook announcement. This contextual cueing or priming of media (Dahlén, 2005) involves implicit activation of memory and forms attitudes towards one stimulus (the message) based on another (the context or media that the message appears in). This means that meetings, announcements, office walls, the internal portal, informal corridor discussions and different people can have a different priming effect. Chose the medium wisely to make sure it is congruent with the message and not preparing the brain for opposite reaction.

- *Symbols.* As Alter (2013) explains, symbols can powerfully shape our thinking and actions because we perceive them effortlessly and rapidly, and because they embed themselves deep in our memories. They are 'magnets of meaning', instantly retrieving associations hardwired in our brains by previous experiences and priming us for specific attitudes, decisions and behaviours. Symbols are culture transmitters (Buchanan and Hunczynsky, 2010). In a much-discussed study, researchers (Fitzsimons *et al*, 2008) primed students by exposing them subliminally to either various Apple logos or various IBM logos. Then they asked them to show creativity by suggesting different uses of mundane everyday items, such as a paperclip (the 'unusual uses' test). Those primed with the Apple logos produced more in number and more creative solutions than the IBM group because, apparently, Apple is wired into the brain as a representation of creative thinking. So, the Apple logo triggers an unconscious reaction that prepares/ primes the brain for creativity more than the Microsoft logo. Such findings strongly suggest that the stimuli we expose ourselves and our colleagues at work to can have a potentially powerful knock-on effect on our behaviour. Take a look around you in the office and at the corporate materials you and your team are using daily. Are there any specific symbols that actually boost your brain's performance? Or are there symbols that can inhibit it? Eliminate the latter and enhance the former. Also, explore which ones can be introduced and how in order to increase the positive impact of symbols on your team's brains.

- *Words*. Priming has first been observed in exercises with associative words, and since then it has been confirmed that it reduces neural processing in the cerebral cortex (Wig *et al*, 2005), meaning that there is less neural activity in the controlled part of our brain than when priming is absent. This is because priming triggers implicit memories that help us perform, mostly automatically, a task based on previous experience without being aware or conscious of this process (Schacter, 1987). Priming saves brain energy and helps us complete tasks efficiently and manage time effectively. Words have increased power in priming and thus we need to pay attention to how we are using them both for our own implicit motivation purposes and those of others in our organizations. In this direction, it makes sense that companies use specifically crafted words for their values that aim to reflect their core philosophy. TOMS Shoes, for example, repeat the word 'give' both in their values and in their corporate materials (TOMS, 2015) to indicate that they are in the business of 'giving' back to communities in need something for every purchase of one of their products (Chu, 2013). There are many lists available of recommended leadership terminology (see, for example, Brandon, 2015) that can induce positive change and improved behaviour, but we suggest that you create your own that fits your role, your aspirations, your organization and your team best. Words will put people, including you, in a specific frame of mind so choose them carefully and strategically.

- *Silence.* More often than not, we have observed a considerable distance between the thinking and the doing worlds in organizations. Office workers are in many cases disconnected with the 'real-world' workers such as factory-level or frontline staff. This kind of disconnect and its risks is what inspired many managers to engage in what is known as 'management by wandering around' (MBWA), a practice aiming to help those in offices to perform reality checks and ensure improvements, especially when actively engaging in problem-solving (Tucker and Singer, 2015). Still, we are often stunned by the amount of talk in offices, in the form of meetings, presentations, reports, etc, and the many times there is an absence of decisive action. An intriguing study by Flegal and Anderson (2008) on the 'verbal overshadowing effect' might explain the reasons for that. They asked high-skilled golfers to first perform an action, then engage in verbal discussions (some explaining their technique, some on irrelevant topics) and at the end to perform the same move again. Describing their experience significantly impaired the higher-skill golfers' ability to re-achieve results. On the contrary, higher-skill golfers who engaged in irrelevant verbal activity did not damage their performance. This proves that the verbal overshadowing effect can actually harm yours and your teams' efforts to repeat great achievements. Too much talk and over-thinking seem to negatively affect expert brains' ability to perform. So, fewer endless explanations and more targeted actions pave the way for better brain preparation for repeating great results.

ACTION BOX: CREATE A MATRIX FOR PRIMING CONDITIONS

Create a matrix table with six rows and two columns. In the first cell of the first column put the title: aspects of priming conditions. In the next five cells of the first column put, one by one, the five aforementioned ways that help creating the right priming conditions, namely culture, context of message, symbols, words and silence. In the first cell of the second column put the title word: organizational elements.

Your target is to fill in the remaining five cells of the second column with those particular aspects that take place in your organization, department or

> team that could influence priming conditions, bringing positive or/and negative results.
>
> Write down your conclusions and consider them the next time that you need to take decisions within your team.

New habits, old habits

We might not be aware of it, naturally, but unthinking routines or habits represent a large number of our everyday behaviours. Studies suggest that as many as 45 per cent of our daily actions are repeated in the same location almost every day (Neal *et al*, 2006). This means that almost half of our behaviour is automated and performed with reduced cognitive interference from our side. This mechanism saves vital brain energy that can be channelled to brain areas and functions where it is needed; if we had to consciously consider and analyse every action we take every single minute then we would be paralyzed by information overload and we would not be able to achieve much. If a habit is beneficial to our leadership potential then we should keep it and even reinforce it, but if it is damaging then we need to reformulate it into something new. Both strategies, though, assume that we know what habits are and how to deal with them, since their massive impact on our behaviour simply cannot be disregarded.

Our neural habit circuits, the brain's regions and connections responsible for habit formation and preservation, have only recently started to be understood. *Reinforcement contingencies* performed by the unconscious mind determine which behaviour will reward us, and which will not, and evaluate the outcome of the chosen behaviour by considering 'reward-prediction error signals' in order to push the formation into a habit or not (Schultz and Romo, 1990). Unconsciously, our brain will nudge us towards behaviour that it perceives as more rewarding, and if this reward actually comes, and it's repeated after we do the same action again and again, then a habit is formed by reinforcement. This is why we might consciously perceive our behaviour as harmful, like eating junk food too often, but we find it so difficult to break the habit. It is the powerful unconscious mind that decides and evaluates the rewards from its own perspective and pushes for a behaviour, not our weaker controlled thinking. When the unconscious mind loves a reward that we consciously – theoretically – believe can be negative, guess who is usually winning. In the brain, the neocortex, basal ganglia, which directs procedural learning, and the midbrain, are associated with habit formation and maintenance, which become more or less engaged as a behaviour is deliberate or habitual (Graybiel and Smith, 2014).

Forming and maintaining a habit is a three-step process:

1 *Exploration.* The neocortex is more active in the first step, when detecting and weighing a new behaviour.

2 *Habit formation.* The basal ganglia become very active as we repeat a behaviour and the feedback loop is reinforced by the reward of the behaviour.

3 *Imprinting.* The habit is well established through neuroplasticity and, surprisingly enough, a part of the neocortex is still active, as if quietly allowing us to engage in the habit. This is the point where there is a possibility to change habits.

Formulating a new habit takes time. In a real-world setting and by using human beings (and not mice, as in many habit-related experiments), it was found that people formed a habit, which means that automaticity settled in, on average in 66 days, with a wide range of 18 to 254 days (Lally *et al*, 2010). This means that habits do not form overnight, and that for neuroplasticity to engrave new settled pathways in the brain, repetition and confirmation of the reward over time are important. Once a habit is formed, new and old goals, expressed by our controlled thinking (conscious) do not interfere much with the process, but for new habits goals are important, both for starting the process and for maintaining it over time until the habit is fully formulated (Wood and Neal, 2007).

Similar to individual habits, there are group habits that can be formulated within a collective entity. In management theory, the relevant concept is of *organizational routines* that are seen as a systematic and repetitive set of activities that occur inside organizations (Feldman, 2000) and that, according to some, are the necessary means of task implementation (Pentland and Rueter, 1994). Routines are organizational entities that produce stability, as well as change whenever they are new, within organizations (Pentland *et al*, 2012). Leaders should consider organizational routines as potentially powerful collective habits that can play a vital role in sustaining and/or changing the status quo within companies.

If you want yourself and your team to develop and sustain new productive habits at the workplace then fully agreeing upon the goal of the habit and ensuring that this goal is emotionally reinforced over time are key steps.

For example, if you want to start keeping reflecting notes with must-do suggestions after every important meeting but you just don't seem to actually be doing it, setting a very strong goal for it and reminding yourself of this goal every time you enter and exit a meeting can help form this new habit. Setting up an automatic reminder to pop up in your e-calendar after meetings, explicitly including the goal and not just the reminded action, has worked well for our clients. Periodically noting down the improvements you managed to do because of this new habit helps cementing it even further through reward reinforcement.

The habit loop is best described by Pulitzer Prize winning reporter Charles Duhigg (2013) in his book *The Power of Habit: Why we do what we do and how to change*. There, he explains the importance of the habit cycle, which goes like this:

1 *Cue.* The trigger that puts our brain in an automatic mode and asks for a specific habit to be retrieved.

2 *Routine.* The physical (doing), mental (thinking) and emotional (feeling) process that is triggered by the cue.

3 *Reward.* The outcome of the routine that the brain craves and the habit exists for.

In order to change an unwanted habit, Duhigg (2013) suggests keeping the cue and rewards the same but choosing a different routine. This is what he calls *the golden rule of habit change*. The essence of this approach is that you can change a habit by trying to achieve the same reward after a specific cue by a different course of action. For example, if you have the habit of automatically accepting more work than your team members in order to achieve the emotional reward of being a 'worthy team member', maybe it is better to help the team deal more efficiently with new workload by facilitating real teamwork and thus achieving the same emotional reward at the end.

Changing the routine in order to reach the same reward is one way of dealing with habits. Working on cues is another (Wood and Neal, 2007). The first strategy is to avoid the cue altogether. If you do not want to be distracted by your impulsive look at new emails or new phone calls, eliminate the cues by turning your phone to silent or switching off email notifications. Similarly, if meetings go automatically sour every time a specific team member teases colleagues negatively, talking personally to this team member to stop providing this cue for habitual collective fighting can stop the undesired behaviour. Whenever cues are environment-specific, changing the environment will cause cues to disappear and thus not trigger the habit anymore. Changing meeting locations, for example, and holding meetings in diverse places within and outside the company has worked miracles in our experience for cutting the habit of valueless prolonged discussions that characterize many business meetings. Changing offices, locations, organizational positions and even jobs can work very effectively in the direction of getting rid of unwanted habits. If the cue cannot be eliminated, like a boss being over-critical in performance evaluation causing a habitual submissive response, then being aware of the cue and the routine can be the start of changing the habit. Inhibiting the cue's automatic response involves the strategy of avoidance, which is about vigilantly monitoring cues and their responses in order to change the course of action. This is a difficult strategy because it involves our conscious mind working hard to change behaviour. It works well if monitoring is really vigilant and assisted by peers. Asking those close to us to help keep an eye on our automated habitual responses and to notify us when a cue appears is crucial for our brains to start moving in a different direction. Training, especially as a structured change intervention, is another form of re-conditioning the brain to respond differently to environmental cues, as is punishment that can take away the pleasure of fulfilling a habit loop.

Entrepreneur James Clear, author of *Atomic Habits: An easy and proven way to build good habits and break bad ones* (2018), highlighted in an interview with *Harvard Business Review* (Beard, 2019) the two main ways of adopting beneficial habits, one being behavioural-driven and the other being cognitive-driven. The behavioural one is his two-minute rule. This rule asks for someone to engage in a small step of a two-minutes behaviour that is in line with the desired habit. Instead of trying to commit to big, often unrealistic goals that require considerable brain energy and a drastic devastation of current habits, break these into smaller achievable and simple steps with a very limited time requirement, and start as soon as possible. The reward of the quick win will lead to the next step, eventually leading to the normalization of the new habit. The cognitive one is about self-image. Instead of focusing on, and stressing about, considerable behavioural changes, try to establish first the kind of person you want to be in relation to this behaviour. He suggests that mentally experiencing and feeling deeply the new image you aspire to achieve, and constantly checking yourself against that preferred model, will create positive behavioural change. In his own words:

The goal's not to run a marathon. The goal is to become a runner. And once you start assigning those new identities to yourself, you're not even really pursuing behavior change anymore. You're just acting in alignment with the type of person you see yourself to be. And so I think in that way, true behavior change is really identity change.

Although quite a bit is already known of how habits are created and work, still there are many secrets to be revealed. Reviewing a large portion of the available neuroscientific literature on the neurobiology of habits, Amaya and Smith (2018) were startled by:

[H]ow something as intuitively simple as a 'habit' exhibits such great complexity when probed scientifically. Recent behavioral neuroscience work has indicated that habits can occur in graded strength, compete with other strategies for control over behavior, are controlled in part moment-to-moment as they occur, and incorporate changes in neural activity across multiple timescales and brain circuits.

They argue that many questions remain unanswered, like the very important one of how exactly positively reinforced habits differ from negatively reinforced habits. Early results indicate that the latter forms faster, but more research is needed to clarify differences in the underlying neuro-chemical processes between the two. In any case, Amaya and Smith highlight the fact that habits can be signs of performance optimization, which is the upside of habits, and of behaviour that is no longer tied to purpose, the environmental changes and to specific goals which, depending on the situation, can be a very serious downsides of habits. Unfortunately, too often we witness managers, teams and even whole departments in companies that have become hostages to habits that maybe initially had strong beneficial reasons to exist but now are just self-serving. Great leaders inspire people to change by providing the right cues, new routines and behavioural reinforcements at the right time.

Habits are major brain automations and need to be understood and utilized appropriately if we are to succeed as managers and leaders. We need to get rid of habits that do not serve our purpose and to develop new ones that do. Desirable habits save valuable brain energy and time, and help us perform effortlessly with greatness. The global popularity of the book *The 7 Habits of Highly Effective People* by Stephen Covey, first published in 1989 and considered to be one of the bestselling business books of all times

(Gandel, 2011), demonstrates that people in organizations want to develop habits that will make them automatically, instinctively, impulsively win. He suggested the following leadership habits:

- be proactive;
- begin with an end in mind;
- put first things first;
- think win–win;
- seek first to understand and then to be understood;
- synergize;
- sharpen the saw.

The neuroscience behind habit formation and maintenance can help us do so, but only if we have a deeper understanding of how and why they work. Which leadership habits do you want to develop? Start today because it will take time. In a true habitual fashion, rewards will be worth the effort.

ACTION BOX: CREATE A HABIT MAP WITH THE SRHI TOOL

The self-report habit index (SHRI) is a tool developed by Verplanken and Orbell (2003) based on measuring key habit features, such as history of repetitions, automaticity and expression identity of the habit. It involves 12 questions to identify the strength of habits. The questions are:

Behaviour X is something that...

1 I do frequently.

2 I do automatically.

3 I do without having to consciously remember.

4 Makes me feel weird if I do not do it.

5 I do without thinking.

6 Would require effort not to do.

7 Belongs to my (daily, weekly, monthly) routine.

8 I start doing before I realize I'm doing it.

9 I would find hard not to do.

10 I have no need to think about doing.

11 Is typically 'me.'

12 I have been doing for a long time.

By scoring each habit on those 12 statements using an agree/disagree scale of 1 to 5 and calculating the average total score, we can have the overall strength of the habit. We use the SRHI to help people and teams in organizations develop a 'habit map'. This is done by following four main steps:

Step 1. Identify three good habits you want to retain and three bad ones you want to change. We recommend you ask your closest people in the organization to help you with this.

Step 2. Complete the SHRI for all habits listed. Ask at least two more people closest to you at work to do the same for your habits, not for theirs. This will give a more independent view on the presence of your habits. Calculate the average scores by combining all responses. Put both the good habits and the bad ones in order of strength (based on their scores), separately. You now have the basis for the map ready.

Step 3. For each habit, good and bad (and even for completely new ones you want to add to the list), create a list of actions to deal with them effectively. Good habits might need solidifying and even further enhancement while bad ones will need toning down and eventually changing. Follow the recommendations in this part of the chapter for doing so.

Step 4. Commit to the process by vigilant monitoring, often team discussion, positive and negative reinforcement whenever possible, and adding/subtracting actions as needed. Habits are sticky and cannot be left to develop or change by themselves.

Have in mind that the 12 questions on the SRHI can be altered to reflect more accurately the specific organizational environment and the nature of your work. Also remember that change is easier when there is peer support than when it is an isolated, individual effort.

You can create a team habit map by repeating the exercise above together with all members of your team. You can then calculate team scores and compare them with personal ones.

In both personal and team maps, repeat the exercise at least six months later, to identify any improvements. If you identify improvements, then celebrate them!

Let's get physical

The world around us is becoming increasingly digital. We spend a big part of our days and nights hooked to the internet both for professional and personal reasons. Consequently, this intensive interaction with the digital world does not leave the brain unchanged. On the contrary, our brain is trying to adapt to the new situation, as it always does when there are significant changes in our external environment. This brain adaptation is, according to some, positive and, according to others, negative. On the positive side, people like Clive Thompson, author of the book *Smarter Than You Think: How technology is changing our minds for the better* (2013), believe that our brain becomes more efficient and effective with the use of new technologies because of advanced collaboration opportunities, easier access to critical information globally and the ability keep track of every aspect of our lives. On the negative side, people like Larry Rosen, author of the book *iDisorder: Understanding our obsession with technology and overcoming its hold on us* (2012) believe that our obsession with mobile devices has already shown signs of pathological brain conditions, like addiction, heightened narcissism, attention deficit and obsessive-compulsive behaviours. These are obviously different sides of the same coin, and it remains to be seen how our brains will benefit, but also suffer, in the long run from these technologies. The fact is, though, that our brains have evolved over millennia though intensive interaction with the physical world and this interaction has created numerous rule-of-thumb types of thinking and reacting in our unconscious mind. The stimuli we expose our senses to every day are signalling to our brain in much more strength than we might realize, influencing our unconscious mind to behave in surprising ways. Being aware of and understanding those real world influences on our powerful concealed thinking can put us in the advantageous position of manipulating them according to our goals. Welcome to the world of 'matter over mind'.

We are our brains, situated within our physical bodies and living in a specific environment that surrounds us constantly. Isolating the one from the other simply does not make sense. The concept of embodied cognition in philosophy and psychology argues convincingly that we cannot separate our brain from its body and the body from its environment. The interrelation of brain, body and the wider biological, psychological and social environment we find ourselves in is critical to how our mind works, and it directly impacts the way we think, feel and behave (Valera *et al*, 1991). According to this approach, most of our cognition is caused from the interplay of our *senso-rimotor* brain regions, which is the neural integration of our sensory and motor systems that help us receive stimuli and move accordingly, with our bodies and the environment we interact with. Evidently, scientists developing artificial intelligence (AI) systems have long recognized that the problem with AI and robotics is not replicating high-level reasoning that actually requires less computation. The challenge is to replicate lower-level *sensori-motor* abilities that need huge amounts of mostly unconscious computations and that are the ones responsible for interacting effectively and efficiently with the world around us in order to survive and thrive (Moravec, 1988). The power of the unconscious is what makes us unique as a species, and we can no longer pretend that it does not really matter or that it mostly damages our supreme intellect. We should know better. Leaders need to know better if they are to improve their teams, their organizations and their societies.

Our mind's dependence on the physical world engulfing it is also vividly reflected in spoken language since we constantly use real-world metaphors to express ourselves on an everyday basis (McGilchrist, 2012). So, we have a 'heavy' schedule this week, a 'clear' decision was made yesterday, we constantly 'push' for reforms, our opponents finally 'dropped' the charges against our firm, the team did not really 'grasp' the importance of the latest announcement, and the new CEO is rather 'cold' compared with the previous one. The list is endless, showing that through evolution our brains have developed a profound ability to deeply consider the natural world around it, even sometimes without separating itself totally from it, in order to always have a realistic view of the situation and react in the best possible way. The fact that a big part of modern thought has been based on the premise that we can somehow be objective and detached observers of the events that surround us is considered by many psychologists, neuroscientists and philosophers alike as not just false but also highly dangerous (McGilchrist, 2012). If we do not interact constantly and actively with the world around us, utilizing all our senses and powerful brain functions, we are in danger of ending up 'living up there, in our heads', isolated from reality, as Sir Ken Robinson accused many academics and modern professionals of doing in 'Do schools kill creativity?', the most-watched TED speech of all time (Robinson, 2006). And this is, unfortunately, what we personally experience in many companies and training sessions we are involved with. There are many people in organizations globally that have been wrongly led to believe that the mind has a life of its own separate from the brain, the body and the environment, and that it presides over all of them. They are too analytical, too dry, too disconnected, and they firmly believe that this is the best option for a true leader. This could not be further from how our brains actually work and what true leaders can do to inspire, perform and grow.

How important is the actual physical environment to our behaviour? Consider the following experimental research results:

- Briefly holding a cup of hot beverage instead of an iced one makes us judge people as having 'warmer, more generous, more caring' personalities (Williams and Bargh, 2008).

- Similarly, if we are asked to hold a hot therapeutic pad instead of a cold one, we are more likely to select a gift for a friend than for ourselves (Williams and Bargh, 2008).

- If we are holding a heavy clipboard instead of a lighter one, we assign higher monetary values to foreign currencies, increase fairness in our decision-making, are more consistent in our deliberations and tend to hold stronger opinions (Jostmann *et al*, 2009).

- In a series of experiments Ackerman and associates (2010) found that sitting on a hard chair will make us negotiate harder, solving a difficult puzzle will make us less socially cooperative after the game, and examining by touch a soft blanket will make us consider someone's personality as more positive than when examining a hard object.

Highlighting the influence of haptic (or touch) sensations for our whole existence, Ackerman *et al* (2010) argue that:

Touch is both the first sense to develop and a critical means of information acquisition and environmental manipulation. Physical touch experiences may create an ontological scaffold for the development of intrapersonal and interpersonal conceptual and metaphorical knowledge, as well as a springboard for the application of this knowledge.

Whenever we present such scientific evidence to our audiences and clients they usually react with great surprise and awe, and a few with rigid scepticism. It is difficult to accept that holding a hot cup of coffee or sitting on a hard chair can really change organizational behaviour within companies and institutions. This shows how much we have removed ourselves mentally from our physical environment, considering our cognitive processes as independent, unaffected and objective. The amount of such research results that indicate the opposite is increasing and inspired psychologist Thalma Lobel to write the book *Sensation: The new science of physical intelligence* (2014), the latter being another name for embodied cognition in the cases when it works to our benefit and does not cause confusion and misguided behaviour. Physical intelligence is actually an excellent term indicating that our brains' ability to take cues from the physical environment in order to move in one direction or another can provide important competitive advantages to those who know how to spot this ability and utilize it. Lobel has included textures, heaviness, colours, distance, brightness, cleanliness, highs and lows, as well as the metaphor-based language we use daily, in her analysis of how our senses influence our unconscious mind and prime us for action. She concludes that:

> Once you become aware of these influences and the power of metaphors, you can use them for your benefit… Pay attention to your senses' input and evaluate it. Your attunement to what your senses tell you will give you physical intelligence – otherwise your senses will yield only data. Armed with this new awareness, you might avoid being swayed by previously unconscious metaphorical associations in your judgments and evaluations of others.

This is the invaluable lesson for embodied cognition; ignoring or denying its existence actually increases the possibility of operating with biases and having a distorted view of reality. This simply means that we can take bad decisions and misleading actions. Accepting it and working with it can increase our effectiveness and ultimately it can become a powerful weapon in your brain-based leadership style. Our recommendation is always for our partners to start feeling the environment around them in a concrete and clearer way. Noticing details of the space around them and searching deep inside to feel how these might change the way they think and act is a good start. We particularly ask them to consider how the physical environment affects their:

- everyday mood;
- repetition of specific chains of thought;
- closeness and connectedness levels to colleagues and other people in the organization;
- creative potential and innovative reflexes.

Even how we form our body postures primes our emotional state in surprising powerful ways. We can fake a posture to produce the desired mental effect. Which means our bodies can literally change our minds (Cuddy, 2012). Remember the mouth-holding pencil exercise that produces happy feelings just by creating a fake smile? The interaction between our unconscious brain, our bodies and the physical environment around us create powerful priming effects on our conscious mind that, if not acknowledged and managed, can lead to considerable mistakes in both thinking and doing. Having said that, evolution has taught this system what works and what is not working so listening to our gut feeling does not mean automatically rejecting it. Actually, gut feelings and intuition play the most significant role for developing leadership expertise for complex and dynamic environments.

Expertise and automaticity

The more we think about a task analytically, the chances are that the less expert we are in this task. Think about it. While we are learning to ride a bike we need to think about all possible aspects of this task in order to avoid falling off. But when we have learned, there is not much thinking going on anymore. We can ride the bike in autopilot and think about something else at the same time. The more you think about a problem, a task or a new idea, the more it shows that expertise has to be developed and not that it is already present. In a seminal study of chess master players, psychologist and chess expert Adriaan de Groot found that chess masters, the ultimate experts of the game, have a remarkable ability 'to quickly capture, and to retain and recall, the information contained in an unknown, complex chess position: its gist and its structure, up to the precise location of almost all men' (Groot and Gobet, 1996). They found that experts processed new information very quickly, and produced an intuitive reaction (through a heuristic or short-cut in the brain) faster than they produced a controlled, conscious thought. Expertise, although it follows complicated pathways in our brains, largely lives in a subcortical structure called cerebellum, or 'the little brain'. This part of the brain has been recently linked to the evolution of our unique human characteristics, such as technical intelligence, since it developed much faster than the neocortex in our ancestors (Barton and Venditti, 2014). It has even, surprisingly, been linked to our creative abilities (Saggar *et al*, 2015). These latest scientific insights show how important it is for our thinking and acting in our everyday working lives. Although older, cortico-centric approaches considered expertise and 'gifted' personal characteristics as a top-down process within the brain, new research suggests that the cerebellum plays a huge role in expertise development, since this is where behavioural refinement takes place (Koziol *et al*, 2010). Traditionally, the cerebellum was thought to be doing this refinement of final action only in relation to motor-related behaviours, but more recently this has been extended to other brain functions involving learning (Doya, 2000). Simply put, 'practice makes perfect' is a cerebellum responsibility and it does not live in the conscious mind. You have to go deeper to find real expertise.

This does not mean, though, that all expertise is fully automatic. On the contrary, thinking and behaving in critical situations when new solutions are needed urgently is a collaborative effort of the conscious and the unconscious minds. Expertise is fine when its fast pattern recognition is applied to rather routine tasks. But when we face chaotic and multifaceted situations, automatic pattern recognition has to be accompanied by fast mental simulations of a few possible outcomes to bring the best results. Pattern recognition alone is highly risky in challenging times, as we saw in Chapter 2. This is the recognition-primed decision (RPD) model developed by Gary Klein (1998), initially based on interviews of fire-fighters' commanders in the USA, as a branch of naturalistic decision-making, which is mostly intuitive and automatic. In crisis situations such as a blazing fire, pattern recognition is very helpful, but could result in the wrong decision because of rapidly changing new conditions, and deliberate analysis alone could be too slow to offer any value at all. So, intuitively retrieving patterns from your long-term memory that you can trust, but at the same time performing a rapid scenario analysis (up to three scenarios) in your conscious mind based on the conditions you see, can be the recipe for leadership expertise in our times.

Our experience shows that influential and inspirational leaders in organizations know how to use their experience as an important decision-making factor, but always contrast this experience and their intuition with current conditions, other people's input and swift scenario analysis. They trust their gut, especially when it sends strong signals towards a decision or behaviour, but they then process it mentally to see where it can lead compared to other possible alternatives. For us, this is clearly the beauty of our brains: nothing happens in isolation and everything is related. This is why we need to try to engage the right parts for the right action, and that is a core aspect of our brain-adaptive leadership model.

BOOST YOUR BRAIN

Try to remember and write down a minimum of three and maximum of five moments from the last year where you demonstrated to your peers a high level of expertise. This could be a vital contribution during a difficult meeting, an answer to a challenging question after a presentation, a winning point in tough negotiations, etc. The focus should be on identifying those heated moments where your expertise proved invaluable for instant problem-solving and when your colleagues were clearly impressed by this demonstration of advanced expertise. For each of these moments create a table with the attributes of your expert intervention. Where did it come from? Experience, intuition, fast pattern recognition, conscious analysis? How much did each of these play a role, and how? By unveiling the hidden forces of expertise in dynamic situations you will be able to understand them better and utilize them more effectively next time you need to save the day in your organization.

Continue this exercise in the future whenever you recognize such a moment again, because this will allow you to include more expert moments and thus help you to build a long-term understanding of your expertise and how to improve it.

Keep in mind

If you separate the mind from the brain and selectively marvel at the supposed supremacy of conscious thinking over intuition and gut feelings you will constantly be disappointed in you and often irritated by those around you. This is because we cannot think and act consciously most of the time and, even more, it is not neurologically possible to 'free' our minds and therefore our decisions and actions from gut feelings and intuition. Actually, it would be dangerous to do so since those provide necessary deep-wired intelligence much needed in our everyday lives as well as free space for our conscious mind to focus on other matters. Adopting your brain's automaticity processes as key elements of what can make you a great leader can drastically improve both your understanding of yourself and others, and remarkably enhance your performance. Using priming to positively and appropriately prepare your brain and the brains in your team, establishing which work habits you need to abolish and which to promote, and creating a physical environment that nudges you and everybody else in the right direction, are crucial for mastering the brain's automatic responses. Becoming a true expert in challenging times requires you to combine this knowledge of how heuristics, intuitions and habits work together with fast, analytical, clear and targeted thinking. Taking into account the unconscious mind's power over our thinking and doing, it all starts from here: what is the readiness potential of your leadership brain?

References

Ackerman, JM, Nocera, CC and Bargh, JA (2010) Incidental haptic sensations influence social judgments and decisions, *Science*, **328**, pp 1712–15

Alter, A (2013) *Drunk Tank Pink: The subconscious forces that shape how we think, feel and behave*, Oneworld Publications, London

Amaya, KA and Smith, KS (2018) Neurobiology of habit formation, *Current Opinion in Behavioral Sciences*, 20, 145–52

Bargh, JA (1994) The four horsemen of automaticity: Awareness, efficiency, intention and control in social cognition, in *Handbook of Social Cognition*, 2nd edn, ed RS Wyer and TK Srull, pp 1–40, Erlbaum, Hillsdale

Bargh, JA (2013) Social psychology cares about causal conscious thought, not free will per se, *British Journal of Social Psychology*, **52**(2), pp 228–30

Bargh, JA (2014) Our unconscious mind, *Scientific American*, 310(1, January), pp 20–27

Bargh, J (2017) *Before You Know It: The unconscious reasons we do what we do*, William Heinemann, London

Bargh, JA, Chen, M and Burrows, L (1996) Automaticity of social behavior: Direct effects of trait construct and stereotype activation on action, *Journal of Personality and Social Psychology*, **71**(2), pp 230–44

Barton, RA and Venditti, C (2014) Rapid evolution of the cerebellum in humans and other great apes, *Current Biology*, **24**(20), pp 2440–44

Baumeister, RF and Bargh, JA (2014) Conscious and unconscious: Toward an integrative understanding of human mental life and action, in *Dual-Process Theories of the Social Mind*, ed JW Sherman, B Gawronski and Y Trope, pp 35–49, Guilford Publications, New York

Beard, A (2019) The right way to form new habits, https://hbr.org/podcast/2019/12/the-right-way-to-form-new-habits (archived at https://perma.cc/5EY4-WZB7)

Bos, MW, Dijksterhuis, A and van Baaren, RB (2006) On making the right choice: The deliberation-without-attention effect, *Science*, **311**(17 February), pp 1005–07

Brandon, J (2015) 40 powerful words to help you lead a team, www.inc.com/
john-brandon/40-words-of-leadership-wisdom.html (archived at https://perma.cc/
3MY5-ELBE)

Buchanan, DA and Hunczynsky, AA (2010) *Organizational Behaviour*, 7th edn,
Prentice Hall, Upper Saddle River, NJ

Cambridge Advanced Learner's Dictionary and Thesaurus, Cambridge University
Press, http://dictionary.cambridge.org/dictionary/english/unconscious (archived
at https://perma.cc/5V3S-YY3V)

Chivers, T (2019) What's next for psychology's embattled field of social priming,
Nature, www.nature.com/articles/d41586-019-03755-2 (archived at
https://perma.cc/8FPS-MWBP)

Chu, J (2013) TOMS sets out to sell a lifestyle, not just shoes, www.fastcompany.
com/3012568/blake-mycoskie-toms (archived at https://perma.cc/
59HV-FQSX)

Clear, J (2018) *Atomic Habits: An easy and proven way to build good habits and
break bad ones*, Random House Business, London

Covey, S (1989) *The 7 Habits of Highly Effective People*, Free Press, New York

Cuddy, A (2012) Your body language may shape who you are, www.ted.com/talks/
amy_cuddy_your_body_language_may_shape_who_you_are (archived at
https://perma.cc/KN9L-HS7S)

Dahlén, M (2005) The medium as a contextual cue: Effects of creative media
choice, *Journal of Advertising*, 34(3), pp 89–98

Doya, K (2000) Complementary roles of basal ganglia and cerebellum in learning
and motor control, *Current Opinion in Neurobiology*, 10(6), pp 732–39

Doyen, S, Klein, O, Pichon, CL and Cleeremans, A (2012) Behavioral priming:
It's all in the mind, but whose mind? *PLOS ONE*, 7(1), e29081

Duhigg, C (2013) *The Power of Habit: Why we do what we do and how to change*,
Random House, London

Feldman, M (2000) Organizational routines as a source of continuous change,
Organization Science, 1, pp 611–29

Fitzsimmons, GM, Chartrand, TL and Fitzsimmons, GJ (2008) Automatic effects of
brand exposure on motivated behavior: How Apple makes you 'Think
Different', *Journal of Consumer Research*, 35(1), pp 21–35

Flegal, KE and Anderson, MC (2008) Overthinking skilled motor performance:
Or why those who teach can't do, *Psychonomic Bulletin and Review*, 15,
pp 927–32

Fried, I, Mukamel, R and Kreiman, G (2011) Internally generated preactivation of
single neurons in human medial frontal cortex predicts volition, *Neuron*, 69(3),
pp 548–62

Gandel, S (2011) *The 7 Habits of Highly Effective People* (1989), by Stephen R
Covey, in The 25 most influential business management books, http://content.

time.com/time/specials/packages/article/0,28804,2086680_2086683_2087685, 00.html (archived at https://perma.cc/E7Y4-5GYF)

Gazzaniga, MS (1998) *The Mind's Past*, University of California Press, Berkeley

Gholipour, B (2019) A famous argument against free will has been debunked, www.theatlantic.com/health/archive/2019/09/free-will-bereitschaftspotential/ 597736/ (archived at https://perma.cc/5VP3-YMDA)

Gigerenzer, G (2007) *Gut Feelings: Short cuts to better decision making*, Penguin, London

Gladwell, M (2005) *Blink: The power of thinking without thinking*, Penguin Books, London

Goldhill, O (2019) Neuroscientists can read brain activity to predict decisions 11 seconds before people act, https://qz.com/1569158/neuroscientists-read-unconscious-brain-activity-to-predict-decisions/ (archived at https://perma.cc/ G8LY-CJE2)

Graybiel, AM and Smith, KS (2014) Good habits, bad habits, *Scientific American*, **310**(6), pp 38–43

Groot, AD de and Gobet, F (1996) *Perception and Memory in Chess: Studies in the heuristics of the professional eye*, Van Gorcum, Assen

Hong, YY, Morris, MW, Chiu, CY and Benet-Martinez, V (2000) Multicultural minds: A dynamic constructivist approach to culture and cognition, *American Psychologist*, **55**(7), pp 709–20

Hood, B (2012) *The Self Illusion: Why there is no 'you' inside your head*, Constable, London

HR Exchange (2018) Employee engagement on the rise: Gallup survey shows increase from 2015, www.hrexchangenetwork.com/employee-engagement/ articles/employee-engagement-on-the-rise-gallup-survey (archived at https://perma.cc/Q3ZF-DUXY)

Jostmann, N, Lakens, D and Schubert, T (2009) Weight as an embodiment of importance, *Psychological Science*, September, doi 101111/j1467-9280200902426x

Kahneman, D (2012) A proposal to deal with questions about priming effects, www.nature.com/news/polopoly_fs/7.6716.1349271308!/suppinfoFile/ Kahneman%20Letter.pdf (archived at https://perma.cc/??????)

Kark, R and Shamir, B (2002) The dual effect of transformational leadership: Priming relational and collective selves and further effects on followers, *Transformational and Charismatic Leadership: The road ahead*, **2**(2), pp 67–91

Klein, GA (1998) *Sources of Power: How people make decisions*, MIT Press, Cambridge, MA

Koenig-Robert, R and Pearson, J (2019) Decoding the contents and strength of imagery before volitional engagement, *Scientific Reports*, **9**(1), p 3504

Kosslyn, SM and Miller, GW (2013) *Top Brain, Bottom Brain: Surprising insights into how you think*, Simon and Schuster, New York

Koziol, LF, Budding, DE and Chidekel, D (2010) Adaptation, expertise and giftedness: Towards an understanding of cortical, subcortical and cerebellar network contributions, *The Cerebellum*, 9(4), pp 499–529

Kristof, ND (2008) What? Me biased? www.nytimes.com/2008/10/30/opinion/30kristof.html (archived at https://perma.cc/YD65-M3A6)

Lally, P, Van Jaarsveld, CH, Potts, HW and Wardle, J (2010) How are habits formed? Modeling habit formation in the real world, *European Journal of Social Psychology*, 40(6), pp 998–1009

Latu, IM, Mast, MS, Lammers, J and Bombari, D (2013) Successful female leaders empower women's behavior in leadership tasks, *Journal of Experimental Social Psychology*, 49(3), pp 444–48

Libet, B, Gleason, CA, Wright, EW and Pearl, DK (1983) Time of conscious intention to act in relation to onset of cerebral activity (readiness-potential): The unconscious initiation of a freely voluntary act, *Brain*, 106(3), pp 623–42

Lobel, T (2014) *Sensation: The new science of physical intelligence*, Icon Books, London

McGilchrist, I (2012) *The Master and His Emissary: The divided brain and the making of the western world*, Yale University Press, New Haven, CT

McGowan, K (2014) The second coming of Sigmund Freud, *Discover*, 24 April, pp 54–61

Meyer, DE and Schvaneveldt, RW (1971) Facilitation in recognizing pairs of words: Evidence of a dependence between retrieval operations, *Journal of Experimental Psychology*, 90, pp 227–34

Moravec, H (1988) *Mind Children: The future of robot and human intelligence*, Harvard University Press, Cambridge, MA

Neal, DT, Wood, W and Quinn, JM (2006) Habits: A repeat performance, *Current Directions in Psychological Science*, 15(4), pp 198–202

Newell, BR, Wong, KY, Cheung, JC and Rakow, T (2009) Think, blink or sleep on it? The impact of modes of thought on complex decision making, *The Quarterly Journal of Experimental Psychology*, 62(4), pp 707–32

Pentland, B and Rueter, H (1994) Organizational routines as grammars of action, *Administrative Science Quarterly*, 39, pp 484–510

Pentland, BT, Feldman, MS, Becker, MC and Liu, P (2012) Dynamics of organizational routines: A generative model, *Journal of Management Studies*, 49, pp 1484–508

Robinson, K (2006) Do schools kill creativity? www.ted.com/talks/sir_ken_robinson_do_schools_kill_creativity (archived at https://perma.cc/58Z5-YHNP)

Rosen, LD (2012) *iDisorder: Understanding our obsession with technology and overcoming its hold on us*, Macmillan. New York

Saggar, M, Quintin, EM, Kienitz, E, Bott, NT, Sun, Z, Hong, WC, Chien NL, Dougherty, RF, Royalty, A, Hawthrone, G and Reiss, AL (2015) Pictionary-

based fMRI paradigm to study the neural correlates of spontaneous
improvisation and figural creativity, *Scientific Reports*, **5**, article 10864

Schacter, DL (1987) Implicit memory: History and current status, *Journal of
Experimental Psychology: Learning, memory and cognition*, **13**, pp 501–18

Schultz, W and Romo, R (1990) Dopamine neurons of the monkey midbrain:
Contingencies of responses to stimuli eliciting immediate behavioral reactions,
Journal of Neurophysiology, **63**(3), pp 607–24

Soon, CS, Brass, M, Heinze, HJ and Haynes, JD (2008) Unconscious determinants
of free decisions in the human brain, *Nature Neuroscience*, **11**(5), pp 543–45

Thompson, C (2013) *Smarter Than You Think: How technology is changing our
minds for the better*, William Collins, London

TOMS (2015) Your impact, https://www.toms.com/impact (archived at
https://perma.cc/DFQ5-AAKS)

Tucker, AL and Singer, SJ (2015) The effectiveness of management-by-walking-
around: A randomized field study, *Production and Operations Management*,
24(2), pp 253–71

Valera, FJ, Thompson, E and Rosch, E (1991) *The Embodied Mind: Cognitive
science and human experience*, MIT Press, Cambridge, MA

Verplanken, B and Orbell, S (2003) Reflections on past behavior: A self-report
index of habit strength, *Journal of Applied Social Psychology*, **33**(6),
pp 1313–30

Wig, GS, Grafton, ST, Demos, KE and Kelley, WM (2005) Reductions in neural
activity underlie behavioral components of repetition priming, *Nature
Neuroscience*, **8**(9), pp 1228–33

Williams, LE and Bargh, JA (2008) Experiencing physical warmth promotes
interpersonal warmth, *Science*, **322**, pp 606–07

Wilson, TD (2002) *Strangers to Ourselves: Discovering the adaptive unconscious*,
Harvard University Press, Cambridge, MA

Winerman, L (2011) Suppressing the 'white bears', *APA Monitor on Psychology*,
42(9), October, p 44

Wood, W and Neal, DT (2007) A new look at habits and the habit-goal interface,
Psychological Review, **114**(4), pp 843–63

Zaltman, G (2003) *How Consumers Think: Essential insights into the mind of the
market*, Harvard Business School Press, Boston, MA

Zimbardo, P (2008) The psychology of evil, www.ted.com/talks/philip_zimbardo_
the_psychology_of_evil (archived at https://perma.cc/JT2H-H78Z)

SUMMARY OF PILLAR 3: BRAIN AUTOMATIONS

Leadership lessons

Understand how priming works

Priming is a process through which we can prepare the brain for specific thoughts and behaviours and it works well when people are not aware of its effect.

Specific leaders' behaviours can prime followers' actions.

Priming can prove very useful as a method of developing leaders.

Five ways to create the right priming conditions in organizations:

- Emphasize cultural values.
- Be aware of the context of messaging.
- Do not neglect symbols.
- Choose the wording.
- Do not underestimate silence.

Maintain or change habits

There is a three-step process of forming and maintaining a habit:

1 *Exploration.* A new behaviour is detected, weighed and accepted.
2 *Habit formation.* Constantly repeat a behaviour that is reinforced by a reward.
3 *Imprinting.* The habit is well established.

You can change unfavourable habits and establish more desired ones by changing the cue and by selecting alternative routines to achieve the same rewards.

Familiarize with the physical environment

The interrelation of brain, body and the wider biological, psychological and social environment is critical to how our mind works, and it influences the way we think, feel and behave as leaders.

Continuously search for how the physical environment affects:

- mood;
- repetition of specific chains of thought;
- closeness and connectedness to colleagues and other people in the organization;
- creative potential and innovative reflexes.

Relations

07

More connected, more successful

SUCCESS, IT'S A LONELY PATH

She didn't trust people. It's not that she couldn't work with them; on the contrary, she had participated in, and even created as a unit director in her current position, effective teams in the various organizations she had worked for. Very early in her career she learned to be cautious, if not suspicious, of her colleagues. She specifically remembers a case where, as a young manager, she was betrayed by another manager she trusted very much. She used to tell him openly about all her doubts and reservations about some of the company's actions and he went on to reveal all these to her director. When she found out about this she felt utterly betrayed and, as recommended by her friends and family, she started trusting no one at work. If she was to make it to the top, it would be a lonely path.

This approach served her well. As she often said to her friends, it is better to be safe than sorry with relationships at work. 'Always keep yourself to yourself and everything will work out fine' was her motto, and consequently she did not seem very open to warm and strong relations with colleagues. While she was approachable, friendly and hands-on with anyone asking her assistance, as soon as people tried to get emotionally closer they hit a wall. She had boundaries that were known to all. She understood that in some cases she probably met people that deserved more from her, but she was not willing to risk anything. 'Better safe than sorry,' she said. That is, until her deputy director left the company and had an exit interview with her. That discussion was the beginning of revising and eventually changing her strategy towards personal relationships at work.

Her deputy was much younger than her but managed to climb up the ladder quickly, and was always considered a young star in the company. The fact that she decided to leave was a blow to the trust-impaired unit director since she was planning for the deputy to replace her as she moved further up the hierarchy. Why did the deputy decide to go? The answer was simple: 'I didn't feel I could get close to you,' she said to the unit director. 'Regardless of the successful cooperation we have had over all these years, I still felt I could not trust you fully and become something more than typical supervisor and deputy, but more like your coach and mentor, with whom you can discuss critical issues,' she continued. 'I want something more from the people I work so close and intensively with,' she concluded. Probing more, the director understood that things were changing at work. Intensive competition both from outside and inside the company, various economic crises around the world, the new socially wired lives we lead, and the unpredictability and dynamism of every day work at the office required a different type of relations. Creative solutions to chaotic and complex new problems asked for closer, deeper and stronger types of relations within and between teams. The 'I am a lone ranger. I trust no one. I succeed by myself' philosophy that probably worked well in earlier decades cannot be applied anymore. We work, create, solve and live together at work in fast rhythms never seen before. Success or failure depends more on the speed and effectiveness of everyday collaboration than on our actual, job-related skills and knowledge. This simple but powerful realization shook the director's belief in 'going at it emotionally alone', and through intensive discussions, personal study and relevant coaching she has now become a master of relations and an expert human collaborator.

Is the best human strategy for success one of cooperation or conflict? Economist and behavioural scientists tried to answer this question through many theoretical constructs, most notably the famous, or infamous depending on how you look at it, Prisoner's Dilemma. Although conceived to prove that humans are better off not collaborating than collaborating, in real life it actually proves the opposite (see box 'The Prisoner's Dilemma for humans' for a detailed explanation). The mindset of a lone ranger or a lone wolf succeeding in all arenas and winning all challenges is an outdated and very dangerous one. Although there might be cases where our individual contribution is more desired and effective than a collective one, such cases are progressively becoming the exception and not the rule. The good news is that our brain is profoundly wired for collaboration and social bonding, and that by understanding the neurological and behavioural underpinnings of cooperation we can reap the benefits faster and better than ever before. Our brains are inherently very good at it; for collaboration to work its miracles, though, we need to want it as well.

THE PRISONER'S DILEMMA FOR HUMANS

The Prisoner's Dilemma represents for us one of the strongest examples of the clash between over-analytical thinking and natural human behaviour or, as some of our students like to call it, of theory vs reality. Although the Prisoner's Dilemma was (and still is) used by some economists as a mental exercise to show that selfish behaviour is preferable to cooperation, it proves exactly the opposite! As the renowned economist and Nobel Laureate Amartya Sen (Nobel Memorial Prize in Economics in 1998) has said, 'the purely economic man is indeed close to being a social moron. Economic theory has been much preoccupied with this rational fool' (Thaler, 2015). And this is more so in the case of the Prisoner's Dilemma.

The game is simple. Two persons have been arrested by the police. They have been separated and offered two options in their interrogation rooms; they can either stay silent (the 'cooperate' strategy) or confess their crime (the 'defect' strategy). The consequences of those choices are as follows (Thaler, 2015):

- If they both remain silent (so both choose to cooperate) they are only charged with a minor crime and leave jail in a year.

- If they both confess (so both choose to defect) they each get five years in jail.

- If one confesses and the other doesn't (one defects and one cooperates) then the confessor is free and the other gets ten years.

What would you do?

We vividly remember the time when, as students, we were first presented with this dilemma. The calculated solution was to defect because the best possible outcome of defecting was better than the best possible outcome of cooperating (free vs one year in jail), and the worse possible outcome of defecting was also better than that of cooperating (five vs ten years in jail). When one of us voiced this calculated opinion in class the economics professor shouted:

'Excellent! This is a prime example of rational thinking. Bravo!' What a victory for economic thinking. The problem is, of course, that by following this strategy we never get the optimum outcome for both, which is one year in jail each. But who cares, since mainstream economics wants humans to be selfish, in the same way that Richard Dawkins' (1976) genes are always acting selfishly to survive within future generations. For mainstream economic thinking, selfishness is the rational way, while cooperation is the primitive, emotional one.

The celebrated Nobel Laureate economist John Nash, his life captured in the 2001 Oscar-winning film *A Beautiful Mind* starring Russell Crowe, mathematically worked out his Nash Unique Equilibrium as a solution to social games like the Prisoner's Dilemma to highlight that the best possible and rational strategy is not to cooperate (Capraro, 2013). However, this is not what happens in reality. Richard H Thaler, one of the fathers of behavioural economics and co-author with CR Sunstein of the 2008 seminal book *Nudge: Improving decisions about health, wealth and happiness*, with a slightly ironic tone towards traditional economic thinkers who often get surprised when encountering real human behaviour, reported in 2015 that:

> The game rhetoric prediction (for the Prisoner's Dilemma) is that both players will defect, no matter what the other player does, it is in the selfish best interest of each player to do so. Yet, when this game is played in the laboratory, 40–50 per cent of the players cooperate, which means [for traditional economic thinkers] that about half the players either do not understand the logic of the game or [for those more open to accept natural human behaviour] feel that cooperating is just the right thing to do, or possibly both.

The Prisoner's Dilemma's main problem is that, as with many concepts in classical economics, it is separated from real conditions. It assumes that people cannot communicate during the process of making a decision, they cannot change their mind, they do not really know each other and they play the game only once. In reality and especially in the business and managerial situations we face daily, all these conditions are actually in reverse (Dimitriadis and Ketikidis, 1999). In many situations, if not in all within our organizations, we know people and they know us, we will most probably continue interacting with them after a specific instance, we can communicate and engage with them while searching for the best strategy and we have reputations that precede us and that will follow us thereafter. Thus, the best strategy is often to cooperate. Even in actual jail-related situations, cooperation seems to be the default strategy and not the exotic option that economists theorized. In an Wired magazine article about two Las Vegas gamblers who managed to illegally hack the algorithm of electronic games in casinos, it was reported than when law officers offered them a version of the Prisoner's Dilemma in order to push them to confess, they actually remained silent and walked free a few months later (Poulsen, 2014). This despite the fact that they had had a very serious argument in the period before, and leading up to, their arrest. Even when the game is played in sterile laboratory environments, many do prefer to cooperate, despite economists' predictions, since we are often inherently optimistic that the best outcome for both players will prevail (Capraro, 2013). The Prisoner's Dilemma for humans suggests collaboration, for *Homo economicus*, defection. In which species do you and your colleagues belong?

On the practical side of choosing the cooperative option as a sustainable leadership strategy, two books had a profound impact on us: Robert Axelrod's classic *The Evolution of Cooperation* (Axelrod, 1984) and the more popular and accessible *The Origins of Virtue: Human instincts and the evolution of cooperation* by Matt Ridley (1997). Axelrod's main contribution was the fact that he organized a competition for software developers to submit specially developed programs and win a Prisoner's Dilemma contest. Those software programs competed with each other simulating prisoners in the Prisoner's Dilemma game. The programs had to include rules of behaviour based on the action and reaction of the other programs. They were asked to engage with each other in multiple rounds of the game each time. He found that the most successful programs that won repeated games were not the ones favouring the defect option, neither the ones blindly cooperating under any circumstances. Based on his findings and suggestions, here are specific strategic steps for modern leaders to maximize the positive effect of cooperation:

- *Be nice:* Never defect first; always start by cooperating. If you defect first, people will take it as your default strategy and they will approach you with extreme caution. In difficult negotiations, in meetings with people you don't know and in interactions with people outside the company, always start by offering your partnership by demonstrating goodwill. In that way you increase the possibility of them cooperating as well.

- *Retaliate/reciprocate:* If they defect, defect too. Do not allow a hostile move to go unnoticed. By defecting when they do, you give a strong message 'I am not a fool; do not take me for granted'. If they cooperate, cooperate too without a second thought.

- *Be forgiving:* Defect when they do, but after that show goodwill again. This is the only way to get out of a vicious and eternal circle of defections that does not produce the optimum outcome for all parties involved. Forgiving and demonstrating a will for better future relations is a winning strategic choice.

- *Communicate:* Reality-based versions of the dilemma occurring in our everyday office lives do not require people to be isolated in detention cells. Actively exchanging reliable information can do miracles for achieving commonly beneficial outcomes. Talk to people, observe their verbal and non-verbal responses, and avoid consciously misleading and tricking your counterparts.

- *Be non-envious:* Focus on maximizing your own score and not necessarily on achieving the best total score in one round. Surprisingly, those in Axelrod's tournament trying to be on top in every round ended up losing the top spot overall. Thus, focusing on personal and not absolute progress helps you to achieve the best results for all.

We have been applying and teaching those principles for many years now, and they have always worked well. We strongly recommend you apply them too, carefully but decisively. You will experience some positive results immediately, but the further you apply them the more long-term the benefits will become.

Matt Ridley (1997), from his side, provided a very comprehensive account of the phenomenon of cooperation, altruism and morality and offered an excellent introduction to the subject. Trying to combine the selfish gene theory with the widespread phenomenon of collaboration, he claimed that 'selfish genes sometimes use selfless individuals to achieve their ends'. He concludes his work by stressing that 'our minds have been built by selfish genes, but they have been built to be social, trustworthy, and cooperative. This is the paradox this [his] book has tried to explain'. The reciprocal strategy of 'I'll scratch your back if you scratch mine' is central to his analysis of why we collaborate outside our kinship and why these collaborations succeed. And it does not have to come from our analytical mind since reciprocity is 'an instinct. We do not need to reason our way to the conclusion that 'one good turn deserves another' (Ridley, 1997). As best put by Axelrod (1984):

> There is a lesson in the fact that simple reciprocity succeeds without doing better than anyone with whom it interacts. It succeeds by eliciting cooperation from others, not by defeating them. We are used to thinking about competitions in which there is only one winner, competitions such as

football or chess. But the world is rarely like that. In a vast range of situations, mutual cooperation can be better for both sides than mutual defection. The key to doing well lies not in overcoming others, but in eliciting their cooperation.

ACTION BOX: THE PRISONER'S DILEMMA IN YOUR TEAM

Separate your team into two groups. Each group should have at least four people. Take the first group and put its members in different rooms: each person must be alone in each room. Do not tell them who exactly their personal partner in the game is. Present each one separately with the Prisoner's Dilemma rules, tell them that they will play it only once, and ask them for their decision: will they cooperate or defect? Repeat the exercise with the second group but this time while all being in the same room, face-to-face with each other and with the ability to play the game three times in a row. Note the decisions of both groups. Discuss with all any possible differences in the strategies between the two groups. Why do those differences exist? Expand the discussion by asking them what the best strategy in the real word usually is. Ask them to refer to real examples from their own experience. You will probably notice that rationality-biased participants will argue hard for the value of defecting, while more natural relationship-builders will be prone to favour the cooperation strategy. Try to build a consensus around Axelrod's steps mentioned in the box 'The Prisoner's Dilemma for humans'.

The socially wired brain

How important are social connections in our lives for who we are, how we behave and what we achieve? Is our brain a personal score-maximization machine or a collective-balancing one? Should we only care about ourselves or should we spend our valuable brain energy on others too? Ultimately, is leadership an individual or a social game? Such questions are central to the BAL model since increasing evidence from neuroscience, psychology and sociology suggest that 'we' is more important than 'I' in surprising and sometimes counterintuitive ways. A leader that cannot understand and connect meaningfully with the brains of others is a leader doomed to fail.

Our brain is a social organ. Social relationships have a dramatic impact on a large number of brain functions and connections, and ignoring them or downgrading them can only harm our overall leadership performance. The mere fact that we have a consciousness, observing and reflecting on our actions and on the surrounding world, is because of our social dimension. Eminent psychologists and neuroscientists attribute the very existence of consciousness to social relationships. Consciousness, for Professors of Psychology Peter Halligan and David Oakley in the UK (Halligan and Oakley, 2015):

> simply occurs too late to affect the outcomes of the mental processes apparently linked to it... We suggest it is the product of our unconscious mind, and provides an evolutionary advantage that developed for the benefit of the social group, not the individual.

This is an amazing statement. The very thing that seems to make us different from each other and even egoistic in our behaviour, the uniqueness of our conscious thinking, is actually a brain function developed by the unconscious to help us survive as a social group, not as an individual. The 'we' created the 'I'! How contradictory this sounds in the light of the profound individualism we observe in our organizations and personal lives. According to Halligan and Oakley (2015), our unconscious mind broadcasts all information and decisions to our conscious mind that then creates a personalized construction necessary for developing adaptive strategies in the real world – such as predicting the behaviour of others, disseminating selected information and being able to adjust perceptions based on external stimuli. And they are not alone in believing so. Neuroscientist Michael Graziano has suggested the attention schema theory to argue that we have our consciousness in order to detect the consciousness of other people and thus to be able to make assumptions about their behaviour. He stated in a recent interview (Guzman, 2015):

> We perceive awareness in other people. And this is crucial to us as social animals. You know, it's not enough to look at another person and think of that person as some kind of robot where you have to predict what the next move of the robot might be. We have this kind of intuitive or gut impression that the other person is aware, has a mind, an inner experience. And we use that to help us understand other people, to help predict their behavior, to interact with them better. And we do this constantly. In fact, we're very much wired to see awareness in other things.

We are wired to use our attention to unconsciously detect signs in the other person's verbal and body language in order to better understand their state of mind and act accordingly. This is why it is called the attention schema theory: we use part of our attention span to pick up patterns in other people's thinking, feeling and doing. Another Professor of Psychology and author of the book *The Neuroscience of Human Relationships: Attachment and the development of the social brain* (2006), Louis Cozolino, emphasized that 'the brain and the body are biological organisms. They are social organisms, such as the neuron in the brain is a social organism. It needs to connect with other neurons' (Sullivan, 2015). If people, and neurons, fail to create adequate connections across the social domain or across the brain respectively, they can be isolated or even rejected from the social and neuro system respectively with probably negative outcomes. As neurons need to create new pathways for the brain to remain healthy and to thrive, people need to create new relationships in order to remain healthy and to thrive. In order for neuroplasticity (as described earlier in the book) to work and to create novel neural pathways, our brain needs to be adequately stimulated. If this does not happen, we have fewer chances for growth. Cozolino concludes, quite rightfully we may add, that (Sullivan, 2015):

> We're only beginning to realize that we're not separate beings, that we're really all just members of a hive, and it may take centuries to make us realize that we're much more interconnected than what we realize now.

Isn't that true for professionals and organizations as well? The more a manager, a company, an industry, even a whole country gets disconnected from its overall environment, the more all these entities become ineffective and eventually decline. Our brain is primarily a connecting, interacting, trusting and cooperating organ and there is the evolution of human kind to provide the testimonial for it. Actually, our species' unique ability to form multi-layered social relations and to collaborate within highly complex and coordinated group activities with genetically unrelated individuals makes the single most important difference for us. Curtis Marean, Director of the Institute of Human Origins at Arizona State University, believes firmly that *Homo sapiens'* extraordinary ability to cooperate, what he calls *hyperprosociality*, is not a learned tendency but a genetically encoded trait and that this is what helped our species dominate against other related species, such as the Neanderthals (Marean, 2015). Although cooperation is also observed in primate species, our unique ability to collaborate in large, well-organized groups by employing a complex morality competence based on reputation and punishment was what gave the edge to humankind (De Waal, 2014).

It might come as a surprise, but what really separates us from other species is not so much our superior reasoning but our social skills. Studies (Tomasello, 2014; Van der Goot *et al*, 2014) have shown that chimps and young human children perform equally on typical IQ tests but human offspring do much better on tests related to social-cognitive skills, like learning from each other (Stix, 2014). So, when someone brags about the power of competition in fuelling personal, economic and social growth, just remind them that without our brain's propensity for cooperation, mutual learning and social fairness we would still be in caves, even totally extinct! As Martin Nowak, Professor of Biology and Mathematics at Harvard University and Director of the Program for Evolutionary Dynamics, put it, 'people tend to think of evolution as a strictly dog-eat-dog struggle for survival. In fact, cooperation has been a driving force in evolution' (Nowak, 2012).

In a similar vein, leadership needs to be seen as a relational phenomenon beyond anything else. As we have just recently argued, true leaders should show care for their followers and each other using their consciousness as a tool for developing in-group collaboration, interdependence and trust as key relational bonds of leadership (Dimitriadis and Psychogios, 2020). The advantages of brain connections between leaders and followers has also been confirmed by a hyper-scanning study which shows that a leader and follower synchronize their neural activities more than when there is a lack of leadership relationship (Jiang *et al*, 2015). Synchronization of neurons can have various positive implications. In a recent study of ours (Psychogios *et al*, 2019) we found that organizational members (both employees and managers) are more connected to each other than their organization, showing a great degree of organizational citizenship behaviour directed to the individual rather than towards their organization. This, though, can be proved extremely beneficial for the organization itself since high levels of commitment to each other lead to high levels of solidarity behaviour that has been proven as a positive organizational outcome.

In other words, leaders in modern organizations cannot go on with a mental state of fierce competition or a feeling of isolation from their social surrounding. On the contrary, wholeheartedly embracing collaboration and cooperation, alongside the necessary competition and unavoidable conflicts, will advance them to true brain-adaptive leaders.

RELATIONS MAPPING IN THE HUMAN BRAIN

Our brain's capacity to create internal spatial maps, or representations, of the physical world in order to navigate effectively is well established in neuroscience. However, the latest research shows that internal spatial mapping, or representation, is used to navigate not just the physical environment around us but the social one as well. Space, grid and border cells in the hippocampus, the brain structure responsible for contextual learning and episodic memory, track those around us by using a specific categorization methodology to help the brain decide how to interact with people (Shaffer and Schiller, 2020). The categorization methodology of these social road maps in the brain 'rates' people in accordance to two variables: power and affiliation. Power refers to the hierarchical position of others in relation to us in social groups. Are we supervisors or subordinates... and to whom? This variable is measured in less or more relative social power. Affiliation is about close or distant social proximity to us. It is based on diverse issues such as biological relatedness, common purpose and goals, and previous experiences when interacting with a person. Is a person a friend or a foe... and when? Power and affiliation comprise our brain's mechanism for mapping social relations and acting accordingly. But it is more than that. It is additional evidence that abstract conscious thinking is a product, or a by-product, of our brains' effort to navigate the physical and social worlds around us. The fact we have a lived, subjective, personal and fleeting experience of internal abstract thoughts in our mind might be nothing more than a side effect of more complex and crucial brain processes of mapping and routing. As Shaffer and Schiller highlight:

> Maps are not accurate portraits of the world in all its complexity. Rather they are representations of relations – distances and directions between locations and what exists where. Maps reduce a dizzying amount of real-world information into a simple, easily readable format that is useful for effective, flexible navigation. [Hippocampal cells]... piece together such related elements into a mental map, which other brain regions can then read out to guide 'navigation', amounting to adaptive decision-making. Mapping allows relations to be inferred, even when they have not been experienced. It also allows for mental shortcuts that go beyond the purview of the spatial and temporal domains. In fact, reasoning using abstract concepts may depend on some of these neural foundations.

Our brain has a stop-and-think capacity, or scientifically put inhibit-and-plan, housed in the executive brain regions, that prevents immediate action and allows abstract modelling of a situation before initiating a corresponding behaviour. This thinking, mind-driven capacity helps us plan better and execute with more success the deeper decisions that are taken by our unconscious brain processes. The value of such mental *and* deeper brain functions is evident when mapping the savannah for prey and hunting for food successfully. Instead of attacking the prey mindlessly on sight and rerouting according to how the prey tries to escape, humans can spend more brain energy on abstractly mentalizing both the best route and possible prey movements in order to hunt it faster and more effectively. The same goes for tribal emigration between jungles, plains and mountains. This mentalization and abstract modelling, though, is applied to social relations too. Instead of behaving reactively and according to inner urges, the human brain recreates an abstract, or incomplete, model of the social terrain and comes up with possible social strategies to navigate successfully within and between groups. This is why thinking exists: not to take a decision, but to map the physical and social worlds and imagine possible routes for a successful application of our deeper decisions and wants. Unconscious motivation is the decision and conscious thinking is the implementation. As inhibit-and-plan was crucial for humans to survive and thrive through hunting, emigrating and avoiding danger in nature, the same goes for surviving and thriving within social opportunities and threats.

I know what you think

The ability of our brain to understand other people's mental states and to use this information to predict their behaviour is really astonishing. It is also an extremely helpful tool for managers and leaders in all organizations around the world. Neuroscience and psychology call this ability theory of mind (ToM) and it represents a core skill of our social brain. The foundations for this theory were laid down in the late 1970s when David Premack and Guy Woodruff wrote their seminar paper 'Does the chimpanzee have a theory of mind?' (Premack and Woodruff, 1978). They explained that:

An individual has a theory of mind if he imputes mental states to himself and others. A system of inferences of this kind is properly viewed as a theory because such states are not directly observable, and the system can be used to make predictions about the behavior of others. As to the mental states the chimpanzee may infer, consider those inferred by our own species, for example, purpose or intention, as well as knowledge, belief, thinking, doubt, guessing, pretending, liking, and so forth.

Simply put, ToM is our species' advanced capacity to understand that we have cognitive, emotive and behavioural processes in our heads and that those are separate from the similar processes in other people's heads. In essence, it is a 'theory' because we can only make inferences of what happens in our and other people's minds since we do not see directly into brains. So, we personally develop 'theories' or 'possible explanations' of what other people think, feel and are about to do. As we explained earlier, the domination of our species versus others depended on our brains' ability to be conscious about our own and other people's actions. Consequently, we can say ToM is the main purpose of having a consciousness.

ToM is a complex system. It combines different brain areas responsible for memory, attention, language, executive functions, emotion processing, empathy and imitation, and it is profoundly dependent on the multi-layered interactions between brain development and social environment (Korkmaz, 2011). Interaction between brain and human environment is crucial. Isolation is fatal. Some children with cognitive and developmental disorders such as autism and schizophrenia have difficulties with ToM (Korkmaz, 2011). Understandably, ToM engages many brain regions in our cortex, such as the prefrontal lobe related to the perception of emotions and the temporal lobe related to recognition of faces, and in deeper structures, such as the amygdala (see for example, Martin and Weisberg, 2003). The fact that the amygdala is involved indicates the guarding role of ToM in our well-being. If the amygdala is hyperactive when engaging in ToM then it has probably detected danger in other people's potential actions and an avoid reaction is triggered. If not, an approach reaction is neurologically allowed and we can proceed to cooperation (you can find more on the amygdala and trust later in this chapter).

The healthy function of ToM is usually tested by specific tasks that reveal if children, or adults (in a greater degree), can recognize correctly that other people can hold a false belief about something. That is, a belief that diverges from reality and can lead to a wrong action. The most famous such test involved two children, Sally and Anne, and was first proposed in early 1980s by Wimmer and Perner (1983). Imagine two children, Sally and Anne, sitting side by side and holding one box each. Sally also holds a stone, which she places into her box and then leaves the room without her box. While Sally is away, Anne opens Sally's box, takes the stone out and places it in her own box. Upon Sally's return, participating children (of various ages each time) are asked to predict in which box Sally will look for her stone. If the answer is in her own box then the children correctly recognized that Sally has a false belief. If they answer that Sally will look into Anne's box then they failed the test as they did not recognize Sally's individuality of mind, a mind that is separate from their own minds, holding opinions separate from reality. Sally does not know that Anne took the stone in her box, as the children themselves know, and thus mistakenly looks into her own box for the stone.

This and other false belief tests in ToM research showcase the importance of mind reading for human understanding, learning and collaborating. Mind reading, not in a supernatural way, but as a very human, brain-based ability to hypothesize what another person is thinking. Shared intentionality, as the famed psychologist Michael Tomasello calls it, is necessary in developing a common ground between members of a group, helping them to achieve joint attention and ultimately to act cooperatively towards a collective goal (Tomasello and Carpenter, 2007). Mind reading, or intention understanding as it is also known, is highly developed in healthy humans and played a vital role in our evolution and dominance as species.

The implications of the ToM for the leadership brain are profound. Leaders play a central role in planning, in overseeing tactical implementation, in conversing, debating and negotiating both internally and externally with a large number of people, and in motivating and directing colleagues and employees. Their ability to understand the state of mind of the people across the table (or even of those absent) and to decode their intentions is key to successful decision-making. Otherwise, leaders are deciding and acting in the dark. But what can we do to improve our mind reading skills? How can we make sure this 'sixth sense' works best for us?

Let's start by abolishing the assumption that adults are by default perfect in mind reading compared to children, and that this ability occurs effortlessly whenever needed. Yes, we do score better than children at tests designed to reveal egocentricity vs perspective taking, but not as dramatically better as expected. A relevant study found that children failed to take the other person's perspective in 80 per cent of the requested tasks, but adults did so in 45 per cent of the tasks (Dumontheil et al, 2010). This is not a spectacular difference. Matthew D Lieberman, Professor in the Departments of Psychology, Psychiatry and Biobehavioral Sciences at the University of

California, Los Angeles, commented on this study, saying (Lieberman, 2013):

> Yes, adults have the capacity to mentalize well, but as this study shows, they don't apply this tendency reliably. This is probably because the brain regions that support accurately mentalizing require effort to work well, and we are wired to be mental couch potatoes whenever we can get away with it. We mentalize a great deal, but this doesn't mean we always do it well or that we can't learn to do it better.

Mentalize is the ability to understand the behaviour of others as a product of their mental state (Frith and Frith, 1999). According to Lieberman, the problem is that too often we allow our mental shortcuts to jump to conclusions about other people's preferences and intention by using as a main input what we ourselves prefer and intend to do. We base our mind reading on our own mental state too much, making serious mistakes in our judgements just because we do not take the time to really consider the situation. He has also found that spending brainpower on non-social mental tasks prior to engaging in ToM lowers our ability to be accurate. On the contrary, the more 'socially' we think before a ToM task, or the more we put our brains in a default mode (relaxing or daydreaming modes), the better we become at it (Lieberman, 2013).

We can personally, and we are sure you can too, find numerous examples from everyday working life when people, including us, fall into those ToM-impairing traps. How many times do we think that we empathize with others or put ourselves 'in someone else's shoes' when in fact we are just projecting our own mental state onto them? Actually, a recent organizational study found exactly that. In a series of experiments with 480 experienced marketing managers, Johannes D Hatula and his associates found that managers who were asked to try to empathize with their customers before predicting what those customers wanted very rarely got it right (Hattula *et al*, 2015). In their attempt to empathize with customers, managers in fact accelerated self-reference because taking the customer perspective activated their own private consumer identity. Their own personal consumption preferences were reflected in the predicted ones. Furthermore, their self-referential preference predictions made them less likely to use market research results, highlighting the fact that wrong mentalization gets even worse with the false confidence that we are good at it. Last but not least, they found that the managers instructed explicitly not to empathize with customers performed better.

Nicholas Epley, Professor of Behavioral Science at the University of Chicago's Booth School of Business, reinforces in his book *Mindwise: Why we misunderstand what others think, believe, feel and want* the point of not trying to empathize with others by lazily fantasizing about their needs. Instead, he suggests revealing other people's actual perspectives by direct interaction with them, or what he calls 'perspective getting' (Epley, 2014). He uses many examples where organizations or individuals understood better what other people wanted by interacting with them and by directly probing their intentions. Such an approach produces much better results than just staying inside an office and assuming other people's needs based on past knowledge, personal experiences and our own preferences. However, he admits that discovering other people's perspectives using a verbal communication approach can be problematic. So, he recommends that when interacting directly with the people you want to mind-read, first try to build rapport so they feel comfortable; second, be as clear and transparent in the discussion as possible; and third, be a good listener and paraphrase often to make sure that you understood correctly what the other person is saying.

Finally, engaging in ToM in groups can be more successful than trying it alone. This means that if you and your colleagues try to mind-read other people (customers, suppliers, people from other departments, etc) as a group, you will probably do it better than if you try to do so by yourself. In a recent study, researchers showed video statements to participants, asking them to spot individually, and in groups, truth or lies. Results indicate that while individuals and groups were equally good at spotting truths, groups were consistently better than individuals at spotting lies; both white, innocent lies and high-stake, intentional ones. According to the researchers (Klein and Epley, 2015):

> This group advantage does not come through the statistical aggregation of individual opinions (a 'wisdom-of-crowds' effect), but instead through the process of group discussion. Groups were not simply maximizing the small amounts of accuracy contained among individual members but were instead creating a unique type of accuracy altogether.

This is the true expression of healthy brain synergy within a team. The whole is greater than the sum, and minds interacting together produce better outcomes than individual and isolated thinking. Isn't this the quintessential goal of teamwork?

The main lessons for the brain-adaptive leader in mind reading are clear:

- Lazy assumptions of what other people think, feel and are likely to do are simply projecting our own state of mind and are doomed to fail.

- Actively trying not to project our own situation upon others helps a lot in taking the right perspective.

- Doing our homework can never be replaced just by mind reading. Market, organizational, industrial and personal research should always be utilized.

- Trying not to engage in tiring mechanical, technical and analytical thinking before engaging in ToM will improve results of mind reading.

- Relaxing, meditating and trying to think about people can improve ToM results.

- Engage directly with others to get their perspective (rather than just guessing it) by creating a comfortable situation, being as clear and transparent as possible, listening actively and making sure you have understood the other person correctly.

- Practise ToM collectively with your team. It can be more successful than doing it alone, especially in spotting deception, which, if unnoticed, can be costly for the organization.

The mirrors in our brains

The concept in the area of mind reading, empathy, imitation and social bonding that has stirred global heated debates in neuroscience and psychology over the last 20 years is that of so-called mirror neurons. Mirror neurons are the neurons that are activated when a human acts as well as when a human observes the same action being performed by another human (Keysers, 2010; Fabbri-Destro and Rizzolatti, 2008; Rizzolatti, and Craighero, 2004).

In 1992, Italian researchers published a study showing that a macaque's neurons fired, not only when the monkey was performing an action but also when this monkey was observing another one doing so (di Pellegrino *et al*, 1992). Since then, numerous studies, opinions and forums appeared either supporting and expanding on the mirror neurons theory or doubting what mirror neurons actually do but not questioning their existence (Marshall, 2014). Celebrated supporters of the wider role that mirror neurons play in our individual lives, group workings and even in our cultures include neuroscientist VS Ramachandran (2011), who believes that the discovery of mirror neurons is equal in significance to the discovery of DNA, and Marco Iacoboni (2009), who claims that mirror neurons are the special brain cells that can finally help us answer centuries-old philosophical and scientific questions about how we successfully organize societies and cultures. Opponents are also quite vocal, with Hickok (2009) identifying a number of empirical problems associated with mirror neurons research, and Kilner (2011) highlighting that the main function attributed to the mirror neurons, action understanding, has alternative neural pathways to materialize itself. Mirror neurons are here to stay and leaders should be aware of their existence and suggested impact regardless of whether this concerns motor skills related functions or deeper emotional ones such as empathy and social bonding.

In the leadership literature arena, Daniel Goleman and Richard Boyatzis published an influential article in *Harvard Business Review* titled 'Social intelligence and the biology of leadership' in 2008 that included a whole section on mirror neurons. For them, mirror neurons play an important role in leadership effectiveness, since followers of the leader will copy or mirror the leader's mental state based on how the leader expresses it. They mention a study where employees presented with negative performance feedback, accompanied by positive emotional signals, reported feeling better overall than employees presented with positive feedback but through negative emotional signals. So, leaders should be very careful about what emotions and mood they show, mainly through their body language, because this profile will be mirrored unconsciously by the people around them (Goleman and Boyatzis, 2008).

In addition to the above evidence, there are some similar studies supporting the idea that mirroring is a significant element of the relational aspect of leadership in particular. First of all, leadership can be perceived as directly related to the phenomenon of social mirroring, since followers bond with leaders by imitating their behaviours (Chartrand and van Baaren, 2009). Among other things, mirror neurons allow leaders to transmit where possible a sense of security. The absence of leadership could be realized subconsciously through the absence of social mirroring, which has been associated with increased cortisol, the stressor hormone (Kouzakova *et al*, 2010). In addition, mirror neurons seem to be associated with another important leadership capability, namely empathy (Cheng *et al*, 2007; Van Overwalle, 2009). By using fMRI tests, Schulte-Rüther *et al* (2007) suggest that mirror neurons are not only involved in motor cognition but also in emotional interpersonal cognition like empathy. In other words, mirroring can help leaders and followers to empathize faster and therefore connect to each other or to a situation.

In a similar vein, mirroring seems to intervene in the way that leaders communicate ideas to followers. A study by Molenberghs *et al* (2017) found that followers found it easier to imagine a leader giving an inspirational talk than they did a leader making uninspiring statements. This study confirmed an example that we share with our students in class. Specifically, we say that someone once said to others, 'I have a dream'. All of us know who he was, don't we? But the point is that we can easily bring this leader from our memory based on his inspirational way of communicating to his followers. We doubt that the same would happen if he had said, 'Guys, we have a problem, let's organize in teams of five to think about it and come up with a solution.' Inspirational ways of communication are definitely enhanced by mirror neurons. In addition, positive leadership experiences seem to be enhanced by the mirror neuron system (Boyatzis *et al*, 2012). In other words, it is highly probable that people will mirror a strong leader's posture, gestures and overall body language as well as the leader's overall mindset. Furthermore, if you find the mindset of your own superiors and/or colleagues incompatible with the purpose of the business then try not to mirror them yourself. Imitation works best when what is imitated is positive to the individual and to the team.

Imitation is a powerful brain mechanism that contributes directly to our development, learning, group assimilation and cultural homogeneity, cooperation and even to innovation, in various stages in our lives. It is a fundamental and hard-wired function of our brains – for example, babies who are only 45 minutes old are able to imitate facial expressions (McGilchrist, 2012). McGilchrist (2012) suggests that imitation, or mimesis, as it is scientifically known:

> let[s] us escape from the confines of our own experience and enter directly
> into the experience of another being: this is the way in which, through human
> consciousness, we bridge the gap, share in what another feels and does, in what
> it is like to be that person... It is founded on empathy and grounded in the
> body. In fact, imitation is a marker for empathy: more empathic people mimic
> the facial expressions of those they are with more than others.

Leaders and managers should not behave as though imitation does not exist and thus ignore their responsibility in shaping other people's moods, attitudes and behaviour. 'Lead by example' is now more relevant than ever due to the amazing attribute of some of our neurons to copy/paste automatically and naturally what happens in front of our eyes. The power of imitation shapes the culture of our organization, departments and teams in ways that go unnoticed to the non brain-adapted managerial mind. We firmly believe, based on our worldwide experience, that mimicking is really the easiest way of shaping a company's or unit's working atmosphere as well as for assimilating desired values in groups. We just need to 'walk the talk', always and in a consistent manner. Mirror neurons and mimesis can work miracles if we choose to acknowledge them and use them in the right direction daily in our working lives.

Human connectivity

Leaders are not, or should not be, islands. They should be actively connected to people inside and outside their organization, with those close to them but also with those further away from their immediate personal circle. We see the creation, maintenance and constant development of powerful human networks as an indispensable aspect of modern leadership in all organizations. This is because our brain, as a primarily *information processing organ* (Sullivan, 2015) needs the right sources and input of information in order for it to create great output in the form of ideas, emotions and actions. Simply put, the quality and quantity of brains your brain interacts with will determine, in a high degree, the quality and quantity of your own brain performance. As a true social animal, the brain-adaptive leader understands the importance of human connections and networks and works tirelessly to build bridges that matter. But which connections are the ones that matter the most in the context of brain calibrated leadership and the complex business and organizational environments we operate in?

The classic social network theory of 'the strength of weak ties' developed by American sociologist Mark Granovetter at the beginning of the 1970s, and advocated strongly since then, is an excellent mental tool for understanding and making better use of our personal connections to other people (Granovetter, 1973, 1983). Granovetter tried to provide a comprehensive explanation of human networks applying simultaneously micro and macro elements of sociological analysis and suggested that we need to urgently look at the ties between people that are not strong, in the sense of relationship proximity (that is, close family members, friends and colleagues). In other words, we need to turn our attention, at least in professional terms, to weak ties, meaning ties with those with whom we have distant and/or indirect connections (that is, acquaintances, distant friends, colleagues and distant family members). The latter had been largely ignored by sociology until 1973. The argument goes that although strong ties are invaluable in providing everyday support, it is the weak ties that inject necessary new ideas, new experiences and new practices in a group of people. Thus, this group must establish links with people that have different characteristics; that is, live and operate in distinct socioeconomic levels. This applies equally to individuals, towns and cities, companies and organizations and even to whole industrial districts with thousands of SMEs collaborating and competing with each other in order to survive in global competitive markets (Dimitriadis, 2008). Fresh ideas, strategic information, out-of-the-box innovations, change initiatives, alternative perspectives and disruptive technologies are rarely the product of tired, over-squeezed and over-familiar relations within groups of people working together too closely for too long. They are born somewhere else in the socioeconomic sphere of human activity and need to be noticed, considered and ultimately embraced by a group of people in order to spawn progress. Think of your organization as a collective brain: if isolated from new stimuli and from the challenges posed by new information, it eventually degenerates and becomes dysfunctional. On the contrary, if it creates and uses appropriately 'weak' links with the outside world in multiple directions then it is kept healthy and neuroplasticity pushes constantly for new pathways, opportunities and capabilities.

In order for weak ties to work best they need to be qualitatively different from a simple 'friend-of-a-friend' connection. If this new link is actively used for producing new information for the individual and the group then it is not just a distant connection but also a very helpful one; it is a bridge (Granovetter, 1983). Based on Granovetter's groundbreaking for its time approach, and on our experience on the topic, we have developed a straightforward matrix to help us explain to leaders and managers the importance of building and maintaining the right bridges. We call it the leadership network connections matrix (see Table 7.1).

TABLE 7.1   Leadership network connections matrix

	Power of connection	
Type of connection	*Dynamic/productive*	*Static/non-productive*
Strong/close	Active operational support	Potential dead end
Weak/distant	Active strategic bridge	Potential beneficial link

Weak ties that are active and productive supply crucial strategic information to the leader and the team, and thus are instrumental in introducing change, innovations and evolution. Leaders should be proactive in creating such connections since this is a core responsibility of the leadership role at any organizational level. For example, they need to attend events such as conferences, conventions and fairs with an open mind and a proactive approach to be able to create new contacts. That's the first step. The second step is to invest personal time to nurture those contacts in order to keep them active and be able acquire from them useful information. Weak ties that exist but are not utilized are potential bridges, and although not much time should be spent on them, they should remain within the leader's mental radar and reach. Concerning strong ties, if they are positively contributing to the leaders' and the team's operational life then they should be maintained and nurtured because they form a crucial support, an everyday network. If they are passive and negative, though, the leader should adopt ways to challenge them and make them reconsider their role within the group and/or team. One of the methods that we usually suggest in our consultancy projects is called 'small and controllable earthquakes'. Leaders need to challenge passive and negative people in order to make them reconsider their attitudes by applying various methods that are critical enough to attract attention, but small in order to be under control. This sometimes requires leaders' personal involvement to guarantee the success of the method. But this is a key role of leaders anyway, to be personally involved. Isn't it?

A brain-adaptive leader cannot afford to be trapped in counter-productive close relations and without distant (socioeconomically speaking), but powerful 'weak' bridges. All those micro and macro relations form our true human network in which we live and work. Their impact on our success or failure and on our emotional and mental states is now revealed to be profound. So profound that even researchers studying human networks for decades were caught by surprise when confronted with the colossal impact of those networks on various human conditions.

THE IMPORTANCE OF FRIENDSHIP

A study by Gallup has shown that having a best friend is a key predictor of superior performance at work, increasing engagement sevenfold (Hallowell, 2011). This is remarkable. Most managers we know around the world would do everything for their team members to increase engagement by 40 per cent or 50 per cent. Seven times, or 700 per cent, increase is unthinkable! The usual managerial toolbox for increasing engagement and productivity, including training, communication and the introduction of new systems, pale in front of the impact of such humble human interventions as nurturing friendships at the workplace. The problem is, as with almost everything concerning management, that training and systems are easily controllable while friendships are not. The comfort of dealing with tools that we can easily budget, monitor and transfer creates an illusion of managerial effectiveness while the things that really matter and can have a considerable impact on performance remain unknown... and even feared. Some managers can resist the idea by claiming that 'we are not here to create friendships but to increase profits'. Such managers need to revise their mindset urgently by immersing themselves into neuroscience to unlock the powerful brain processes that can increase productivity and profits more than any traditional managerial tool.

Friendship has multiple evolutionary benefits. It helps us achieve optimum health and well-being mainly by allowing us to deal effectively with distress, uncertainty and threats (Pattee, 2019). It creates a social support system with people that 'get us' and that 'we get', meaning that we understand each other better and faster, and we connect in an emotionally deeper way. The emotional and health benefits of friendship appear when this relationship is a close rather than a distant one. And this is the main challenge of friendship: close friendships are difficult to build and maintain. Evolution created a complex process of allowing friends to be close to us in order to make sure someone that appeared initially as a good friend does not turn out to be an enemy. Creating close friendships is crucial both for mental health and for productivity in the workspace. Here are few recommendations for how to create them (Pattee, 2019):

1 *Nurture a sense of security.* Closeness requires safety by avoiding foes in our inner circle that could do a lot of emotional damage if not detected and

rejected early. Security requires the following behavioural traits: consistency, availability, reliability, responsiveness and predictability.

2   *Active attentiveness.* Actively looking at and listening to your friend by focusing in that moment of social sharing is crucial for developing closeness.

3   *Be vulnerable.* Sharing information that makes you look like a normal human being, with both strengths and weaknesses, forges stronger friendships. Gradually let your guard down and show your true, authentic self to those that you want to develop a close friendship with.

4   *Testing.* The developmental process of trust requires small steps of reciprocal behaviours that constantly confirm the friendship and lead to a bigger, more meaningful next step. Asking small favours and being ready to reciprocate are the key testing elements of friendships.

5   *Individualize friendships.* There is no one-size-fits-all approach to close friendships. Different people have different emotional needs, leading to complex, dynamic and differentiated social ties that always depend on the people involved.

In a ground-breaking study in 2018, Parkinson and associates showed that the brains of close friends react in a similar way to external stimuli. They asked people to view naturalistic movies while in an fMRI having their brains scanned for reactions to the movies. Those people that were close friends had similar brain reactions to the movies, while the more the social distance increased, the lower were the similarities of their brain reactions. According to the researchers (Parkinson *et al*, 2018): '[t]hese results suggest that we are exceptionally similar to our friends in how we perceive and respond to the world around us, which has implications for interpersonal influence and attraction'. Having close friends at work not only increases resilience and willpower, it also creates a common perception of how the world behaves and what we need to do about it, which leads to synchronized thinking and coordinated behaviours. Thus, engagement and productivity increase.

Nicholas Christakis and James Fowler, in their highly appraised book *Connected: The surprising power of our social networks and how they shape our lives*, took the world by storm when they revealed the unexpected ways in which human networks influence who we are. Their main conclusions

(Christakis and Fowler, 2010), summarized in three groups here, are discussed through the lenses of the BAL framework:

1 *We make networks.* We constantly, consciously or unconsciously, define our participation and position in human networks. To how many people are we going to be connected, strongly or weakly, and what role are we going to play in these relationships? Such actions, either as deliberate decisions or intuitive behaviours, will determine the experience that we will have in the network. We do not always choose the people we work with but we choose the people we are really connected with. Thus, leaders should never hide. They should purposely acquire central and active positions in multi-layered networks inside and outside their organizations, connecting to people in accordance with the leadership network connection matrix discussed above. Leaders make networks, they do not simply participate in them.

2 *Networks make us.* The number and quality of connections determines our own state of mind, well-being, habit formation and happiness levels. And this goes beyond the people we know. It also involves the people that are connected with the people we know and that we have never met. Transitivity, the process of characteristics passing through different degrees of separation from people to people (through imitation) works for up to three degrees of separation. This means that the friend of a friend of a friend of yours, who you have never met, influences your happiness and your health habits more than you ever thought was possible. In other words, a decision that a person will make with whom you are indirectly linked (practically meaning, you do not know him/her personally) can influence the decision of a person that you know, that in turn can influence a decision of yours. Once again, here is chaos theory in action, since this works best in complex and rich networks and not in connections arranged as a simple line. Social reinforcement from multiple connections is important for norm and behavioural change. Leaders therefore need to be cautious about the portion of their behaviour attributed to network transitivity. Thus, observe and monitor behavioural patterns in your team and explore their origin. If positive, reinforce them. If negative, try to disassociate through increased awareness, meaning that noticing them will most probably make you alter your behaviour accordingly. Not noticing them will allow the situation to persist. On the other hand, to create desired norms a leader should take advantage of transitivity and increase interactions that promote those norms. This is possible

by introducing more people that behave accordingly in different levels of the organization and even from the outside of the company. Leaders are aware of network influences and use them to their advantage; they do not just blindly go along with them.

3 *Networks are alive (and often independent of us).* Complex, direct and indirect relations in dense human networks have collective characteristics that are greater than the sum of their parts. Networks exist and change independently from each individual contribution and remain active before and after each individual participation. However, they also have emergent properties which are attributes deriving from the multi-layered interactions and interconnections of members. In that way, cultures depend less on individual actors and more on the interplay of many actors that will consciously or unconsciously interact with its norms and with each other. Leaders of larger organizations are in the position to influence their culture in a slower manner and by using more tactics than those in a smaller one. In both cases, though, leaders should be aware of the fact that cultures are complex living entities that do not change by simply changing a rule or a person. They move in unison. In this respect, change agents, opinion leaders and influencers in general cannot achieve great results if previously people do not have the time to adjust in the new situation. In other words, a change, especially a radical one, cannot be applied on the spot (at its emergence), but after a period of time when people become more familiar and aware of it. This is why dramatic changes in opinions, attitudes and behaviours within a culture rarely happen without first making sure that people collectively have reached a critical point of accepting those changes (Watts and Dodds, 2007). Leaders orchestrate changes; they do not necessarily bring them about alone.

In a hyper-connected environment created by social media and the online world, Christakis and Fowler (2010) believed that social networking changes in the digital domain but also that its fundamentals remain the same. It is true that Facebook, LinkedIn, Twitter, Instagram and other social media allow us to make more connections faster and easier than ever before. We can never use avatars to hide our true identity. But are all those connections active support networks and real bridges? We can't say for sure because it depends on the connection. Social media have accelerated connections and often provide real value in transitivity. But they do not automatically mean active support or bridges. As Dr Jillian Ney, a social intelligence expert, convincingly argued in her 2014 TEDx speech 'Connectedness and the digital self', we need to put connectedness, a deeper psychological link, back into the discussion of online connections (Ney, 2014). We, too, believe that being connected to people or organizations through social media does not necessarily mean engaging mentally and emotionally with them and thus cannot be considered *a priori* as active human network connections. Undoubtedly, leaders and managers need to be present in social media (and we will discuss in the next chapter how this should be done) and create potentially active links, but they still need to nurture and deliberately manage connections to people online and offline in order to experience positive and dynamic relations with them.

Ultimately, we see ourselves, as Christakis and Fowler (2010) do, as members of super-organisms: the living, breathing human networks we belong to. As such, we ought to contribute decisively to them, to shape them but also to be aware of how they shape us. Our brain leadership potential depends on that.

THE ART OF HUMAN CONNECTIVITY FOR LEADERS

How can modern leaders become better at connecting with other people inside and outside their organizations? Ori and Rom Brafman, suggest that in order to accelerate our ability to create instant deep connectivity, something they call 'clicking' with other people, we need to increase our vulnerability (Brafman and Brafman, 2010). Although this may sound counterintuitive, we can connect instantly to the deeper emotional and mental states of people if we allow our-selves to look open and vulnerable. In that sense, vulnerability induces instant trust and instincts of care in other people. Of course, this has to be expressed appropriately because over-vulnerability can give unwanted weakness signs. Brafman and Brafman (2010) refer to a vulnerability continuum present in human interaction. At one end we have non-vulnerability and at the other high vulnerability. The five distinct stages on this continuum, as revealed though everyday conversations, are as follows:

- *Phatic stage.* We speak with social niceties, words void of any real emotional charge like 'How are you?' and 'Nice to see you.' The purpose here is not to extract any specific response but to decrease any possible problems in our interaction.

- *Factual stage.* We speak with actual facts and data that do not elicit any subjective opinions, like 'I am an engineer' or 'I live in Europe.' Such straightforward observations reveal a bit more about us but not in a deep emotional way.

- *Evaluative stage.* We give personal views or opinions about people, places and situations like 'I like this plan' and 'This product is amazing.' Here we take a limited social risk because other people might not agree. Still nothing dramatic though.

- *Gut-level stage.* We speak in emotionally revealing statements like 'I am so sad Mark is leaving the company' or 'I am really glad I hired you.' Those statements are more risky and leave us open to criticism and to emotional disagreement. However, they are the kind of statements made to people closer to us, people in our circle of trust.

- *Peak stage.* This is the most revealing stage, where we open our heart and mind, and speak out our inner feelings, deep fears and wildest hopes. Statements like 'I really wonder how we are going to make it this month with so low numbers. I fear for my job, although it is inconceivable to me that I am in danger, having spent my whole life in this office. Do they not care about the past? Am I not important to them beyond numbers? This is scary...'

The first three stages are more transactional, while the last two are more connective. It is our duty to assess a situation and decide which approach we have to employ in order to connect faster and deeper with the people we talk to. Starting at stage 5 with someone we just met is not advised. But moving gradually from the first three to the last two, even during a single conversation, can help us show some degree of vulnerability and connect instantly with people.

The way by which leaders can mould their behaviour in order to become more connectable to others in the long run, based on neuroscientific and psychological findings, is described in the bestselling book *The Charisma Myth: Master the art of personal magnetism* by Olivia Fox Cabane (2013). She summarized the characteristics of charismatic leaders into three clusters. For us, these clusters correspond perfectly with the 'three brains' approach presented earlier in our book:

- *Cluster A: Presence.* Portraying the feeling that you are 100 per cent focused on the issue and the person (or persons) in front of you. We have all experienced situations in our jobs where we felt that the person in front of us was not actually following or listening to what we were saying. Not being focused makes other people feel distant and even betrayed. Great leaders make people feel noticed and important, regardless of their position. That's absolutely necessary for creating meaningful connections inside and outside the company. This cluster refers to the older, more primitive brain because it is through our body language, posture, gaze and tone of voice that people will understand, consciously or unconsciously, that we are there for them. The automatic brain will pick up signals portraying presence or not, even when you talk to people over the phone.

- *Cluster B: Warmth.* Empathizing with people and showing that you really care about their situation is the second cluster. Goodwill, compassion and acceptance are core processes here, since through warmth we make people feel comfortable, looked-after and protected. This refers to the emotional brain, and especially the mid-brain systems of approach and avoidance, as discussed in previous chapters, since our own stance will make people feel safe and get closer to us or feel threatened and get away from us. Great leaders give strong and clear signs of compassion and thus create meaningful connections that last.

- *Cluster C: Strength.* This is related to perceptions of power and authority and, although it does include some elements of the primitive and mid brains, it is mostly rooted in the neocortex. It refers to our ability to do something specific for the people we want to connect with. Think about it. Being present is great and being compassionate is even greater. But real leadership requires you to be able to take specific decisions and actions that will practically improve your team's performance. Your ability and willingness to demonstrate in action that you can help is a matter of advanced analytical and technical skills. Nobody can connect to a leader that is unwilling or incapable of demonstrating (proving) her or his expertise when most needed.

New voices are added on the importance of warmth and strength both in our personal lives (Neffinger and Kohut, 2014) and in brand management (Malone and Fiske, 2013) since these two variables actually represent affiliation and power, as explained in the social mapping section earlier. Executive presence has also become an area with its own advocates and expanded scope (see for example Hewlett, 2014). Although we can rarely be perfect in all three, we believe that actively noticing our performance in all of them will help us improve the way we connect and maintain relations.

Trust the amygdala and your brain chemicals

Trust and the amygdala do not usually go hand in hand. This is because whenever there is heightened activity in the amygdala we trust less and when the amygdala is quiet we trust more. We do, though, need to trust, in the sense of listening to, the amygdala on this.

The amygdala is usually considered to be the culprit when people over-react with fear and anger in a situation, and thus fail to manage their emotions effectively and deeply connect with others (see Chapters 3 and 4). However, the amygdala plays a crucial role in regulating the level of trust-worthiness of other people. Simply put, without the amygdala we would trust everyone without any limits and with catastrophic results for us and our organizations (Fox, 2013). The amygdala, together with the insula and other brain centres, is a key neural area activating our alarm system when confronted with untrustworthy-looking faces (Winston *et al*, 2002). This hard-wired cautionary system, provided by the amygdala to protect us in risky situations, is absent if the amygdala is dysfunctional. Actually, people with a deactivated amygdala have their natural guard down and are ready to trust anyone with everything (Adolphs *et al*, 2002). It is the amygdala that fires up in order to detect fear and anger in other people and even to differentiate between those two emotions (Fox, 2013). Since faces express-ing those two emotions are deemed usually as less trustworthy, the absence of a healthy amygdala stops us from taking precautionary measures when interacting with them, making people without an amygdala even ready to give their credit card details out to random strangers if asked to do so (Fox, 2013). This is because the amygdala has an important role in evaluating the emotional valence of stimuli. For example, people with amygdala injuries have a difficulty in learning associations between environmental stimuli and emotional states. Thus, they may fail to learn that a stimulus predicts reward or danger; they may also fall in social rank, or show decreased affiliative behaviour (Blair, 2001). Thus, we should be extremely grateful for a fully functioning amygdala in separating out which people to trust and not to trust. If you experience feelings of avoidance when meeting people, explore those feelings before accepting or rejecting them. It might be your healthy amygdala sending you justified signs of caution or it could be an oversensi-tivity of your avoidance system making you less approachable. Learning is a core skill of brain-adaptive leaders.

In the opening case of this chapter, our protagonist had nurtured a hyper-active amygdala that did not allow her to connect meaningfully to the most important people around her at work. In our modern, dynamic and complex global environment, more and more employees choose to work for companies they feel personally and emotionally connected to, as we saw from the 2015 results of *Fortune*'s top 100 best companies to work for in Chapter 5.

Apart from the amygdala, there are two important brain chemicals involved in developing trust between ourselves and people in our human network. One is oxytocin and the second is vasopressin. The latest studies suggest that oxytocin plays a central role in organizing our social behaviour because it increases our propensity to trust others, approach and connect with them, and create meaningful relationships (Levitin, 2015). We can significantly increase the flow of oxytocin by playing and enjoying music together with other people (ibid.) or even by simply giving them a big hug eight times per day (Zak, 2011). The second neurochemical is called vasopressin and is crucial for regulating bonding, sociability and stress responses between humans and other mammals (Levitin, 2015). This chemical has been found to contribute extensively to retaining relationships by decreasing the defect option (Young, 2003). It increases close bonding between people by attaching excitement to the relationship, but it can wear out over time. Thus the challenge is to identify ways to reinvigorate the relationship when we detect signs of emotional detachment in others (Young, 2003). The leader's role is to keep their connections alive, dynamic and productive, and this means infusing excitement and positive surprises in their relations to be able to maintain workable levels of neurochemicals facilitating bonding.

The exclusion of humans at work, in terms of machines replacing humans in many typical job positions, has been prophesied by many but most notably by Jeremy Rifkin in his classic book *The End of Work: The decline of the global labor force and the dawn of the post-market era* (1995). More recently, the *Harvard Business Review* in its June 2015 issue included a front page with the picture of a robot and the title 'Meet your new employee: How to manage the man–machine collaboration'. Although machines play an increasing role in many areas of business and organizational activity, the centre stage is still set for humans. In his article 'Humans are underrated' in *Fortune* magazine, adapted from his book of the same title, Geoff Colvin explains that humans will remain in charge for the foreseeable future, as only they can satisfy the deep interpersonal needs that are so profoundly important for engagement inside and outside our organizations (Colvin, 2015). The fact that empathy levels, as measured by 72 studies of 14,000 students in the US, have fallen by more than 10 per cent in the last 30 years, shows that although empathy is in great demand ('empathy is the critical

21st-century skill' for Meg Bear, the Group VP for Oracle), its supply is shrinking and this requires a dramatically new turn in how we prepare workers, managers and leaders today (Colvin, 2015). Colvin concludes, and we couldn't agree more, that:

> For the past 10 generations in the developed world, and shorter but still substantial periods in many emerging markets, most people have succeeded by learning to do machine work better than machines could do it. Now that era is ending. Machines are increasingly doing such work better than we ever could... Fear not... you'll find that what you need next has been there all along. It has been there forever. In the deepest possible sense, you've already got what it takes. Make of it what you will.

BOOST YOUR BRAIN

Ask your team to write down all the necessary skills to succeed in your goals. Rate them from 1 to 5 based on how effectively they can be done by machines (closer to 1) or by humans (closer to 5). Calculate the total scores and averages for all skills, and use the results to initiate a group discussion about the unique characteristics of human contribution. You can also discuss how machines can assist, but the main purpose is to identify what makes humans capable of creating an engaging environment and developing social connections that matter. How can the team improve? Use the materials in this and previous chapters to progress the discussion and to build consensus.

Keep in mind

Leaders are not islands. On the contrary, they take central positions in organizational and business networks in order to facilitate relations that will help them achieve their goals. However, they need to recalibrate their brains to better understand the social behaviours of themselves and others, and to constantly connect deeper with people both inside and outside their organizations. Leaders need relations mapping to fully appreciate the socially based origins of consciousness, adopting a 'collaborate first and then reciprocate' attitude in business interactions, utilizing imitation as a cultural creation and maintenance tool, establishing and managing extensive human networks, managing the art of clicking and of demonstrating presence. Also, they need to introduce the concept of friendship and based on warmth and strength, make sure that the amygdala and some brain chemicals are in a careful collaborative mode; these are necessary conditions for adapting our brain for networked leadership. The era of the machines also paradoxically bring the era of human beings. Be the first to thrive by cultivating in your brain what makes us uniquely human!

References

Adolphs, R, Baron-Cohen, S and Tranel, D (2002) Impaired recognition of social emotions following amygdala damage, *Journal of Cognitive Neuroscience*, **14**(8), pp 1264–74

Axelrod, R (1984) *The Evolution of Cooperation*, Basic Books, New York

Blair, HT (2001) Synaptic plasticity in the lateral amygdala: A cellular hypothesis of fear conditioning, *Learning and Memory*, **8**(5), pp 229–42

Boyatzis, RE, Passarelli, AM, Koenig, K, Lowe, M, Mathew, B, Stoller, JK and Phillips, M (2012) Examination of the neural substrates activated in memories of experiences with resonant and dissonant leaders, *The Leadership Quarterly*, **23**(2), pp 259–72

Brafman, O and Brafman, R (2010) *Click: The forces behind how we fully engage with people, work and everything we do*, Crown Business, New York

Cabane, OF (2013) *The Charisma Myth: Master the art of personal magnetism*, Portfolio Penguin, London

Capraro, V (2013) A model of human cooperation in social dilemmas, *PLOS ONE*, **8**(8), e72427

Chartrand, TL and van Baaren, RB (2009) Human mimicry, *Advances in Experimental Social Psychology*, **41**, pp 219–74

Cheng, Y, Meltzoff, AN and Decety, J (2007) Motivation modulates the activity of the human mirror neuron system, *Cerebral Cortex*, **17**, pp 1979–86

Christakis, NA and Fowler, JH (2010) *Connected: The surprising power of our social networks and how they shape our lives*, HarperPress, London

Colvin, G (2015) Humans are underrated, *Fortune*, European Edition, **172**(2), 1 August, pp 34–43

Cozolino, L (2006) *The Neuroscience of Human Relationships: Attachment and the development of the social brain*, WW Norton & Company, London

Dawkins, R (1976) *The Selfish Gene*, Oxford University Press, Oxford

De Waal, F (2014) One for all, *Scientific American*, Special evolution issue: How we became human, **311**(3), September, pp 52–55 di Pellegrino, G, Fadiga, L, Fogassi, L, Gallese, V and Rizzolatti, G (1992) Understanding motor events: A neurophysiological study, *Experimental Brain Research*, **91**(1), pp 176–80

Dimitriadis, N (2008) Information flow and global competitiveness of industrial districts: Lessons learned from Kastoria's fur district in Greece, in *Innovation Networks and Knowledge Clusters: Findings and insights from the US, EU and Japan*, ed EG Carayannis, D Assimakopoulos and M Kondo, pp 186–209, Palgrave Macmillan, London

Dimitriadis, N and Ketikidis, P (1999) Logistics and strategic enterprise networks: Cooperation as a source of competitive advantage, 4th Hellenic Logistics Conference, Sole – The International Society of Logistics – Athens Chapter (in Greek)

Dimitriadis, N and Psychogios, A (2020) Social brain-constructed relational leadership: A neuroscience view of the leader–follower duality, CAFÉ Working Papers 1, Centre for Applied Finance and Economics, Birmingham City Business School

Dumontheil, I, Apperly, IA and Blakemore, SJ (2010) Online usage of theory of mind continues to develop in late adolescence, *Developmental Science*, **13**(2), pp 331–38

Epley, N (2014) *Mindwise: Why we misunderstand what others think, believe, feel and want*, Allen Lane, London

Fabbri-Destro, M and Rizzolatti, G (2008) Mirror neurons and mirror systems in monkeys and humans, *Physiology*, **23**, pp 171–79

Fox, E (2013) *Rainy Brain, Sunny Brain: How to retrain your brain to overcome pessimism and achieve a more positive outlook*, Arrow Books, London

Frith, CD and Frith, U (1999) Interacting minds: A biological basis, *Science*, **286**(5445), pp 1692–95

Goleman, D and Boyatzis, R (2008) Social intelligence and the biology of leadership, *Harvard Business Review*, **86**(9), pp 74–81

Granovetter, MS (1973) The strength of weak ties, *American Journal of Sociology*, **78**(6), pp 1360–80

Granovetter, M (1983) The strength of weak ties: A network theory revisited, *Sociological Theory*, **1**(1), pp 201–33

Guzman, IP (2015) What is consciousness? Dr Michael Graziano and the attention schema theory, *BrainWorld*, **6**(2), Winter, pp 46–49

Halligan, P and Oakley, D (2015) Consciousness isn't all about you, you know, *New Scientist*, **227**(3034), 15 August, pp 26–27

Hallowell, EM (2011) *Shine: Using brain science to get the best from your people*, Harvard Business Review Press, Boston, MA

Hattula, JD, Herzog, W, Dahl, DW and Reinecke, S (2015) Managerial empathy facilitates egocentric predictions of consumer preferences, *Journal of Marketing Research*, **52**(2), pp 235–52

Hewlett, SA (2014) *Executive Presence: The missing link between merit and success*, HarperCollins, New York

Hickok, G (2009) Eight problems for the mirror neuron theory of action: Understanding in monkeys and humans, *Journal of Cognitive Neuroscience*, 21(7), pp 1229–43

Iacoboni, M (2009) *Mirroring People: The science of empathy and how we connect with others*, Picardo, New York

Jiang, J, Chen, C, Dai, B, Shi, G, Ding, G, Liu, L and Lu, C (2015) Leader emergence through interpersonal neural synchronization, *Proceedings of the National Academy of Sciences*, 112(14), pp 4274–79

Keysers, C, (2010) Mirror neurons, *Current Biology,* 19(21), pp 971–73

Kilner, JM (2011) More than one pathway to action understanding, *Trends in Cognitive Sciences*, 15(8), pp 352–57

Klein, N and Epley, N (2015) Group discussion improves lie detection, *Proceedings of the National Academy of Sciences*, 112(24), pp 7460–65

Korkmaz, B (2011) Theory of mind and neurodevelopmental disorders of childhood, *Pediatric Research*, 69, pp 101R–108R

Kouzakova, M, van Baaren, R and van Knippenberg, A (2010) Lack of behavioral imitation in human interactions enhances salivary cortisol levels, *Hormones and Behavior*, 57, pp 421–26

Levitin, DJ (2015) Why the modern world is bad for your brain, *Guardian*, www.theguardian.com/science/2015/jan/18/modern-world-bad-for-brain-daniel-j-levitin-organized-mind-information-overload (archived at https://perma.cc/7Q46-7LPJ)

Lieberman, MD (2013) *Social: Why our brains are wired to connect*, Broadway Books, New York

Malone, C and Fiske, ST (2013) *The Human Brand: How we relate to people, products and companies*, Jossey-Bass, San Francisco

Marean, WM (2015) The most invasive species of all, *Scientific American*, 313(2), August, pp 22–29

Marshall, J (2014) Mirror neurons, *Proceedings of the National Academy of Science USA*, 111(18), p 6531

Martin, A and Weisberg, J (2003) Neural foundations for understanding social and mechanical concepts, *Cognitive Neuropsychology*, 20(3–6), pp 575–87

McGilchrist, I (2012) *The Master and His Emissary: The divided brain and the making of the western world*, Yale University Press

Molenberghs, P, Prochilo, G, Steffens, NK, Zacher, H and Haslam, SA (2017) The neuroscience of inspirational leadership: The importance of collective-oriented language and shared group membership, *Journal of Management*, 43(7), pp 2168–94

Neffinger, J and Kohut, M (2014) *Compelling People: The hidden qualities that make us influential*, Piaktus, London

Ney, J (2014) Connectedness and the digital self: Jillian Ney at TEDx University of Glasgow, www.youtube.com/watch?v=3QA8iy7sjT8 (archived at https://perma.cc/K8A6-JTFY)

Nowak, MA (2012) Why we help, *Scientific American*, **307**(1), July, pp 20–25

Parkinson, C, Kleinbaum, AM and Wheatley, T (2018) Similar neural responses predict friendship, *Nature Communications*, **9**(1), pp 1–14

Pattee, E (2019) How to have closer friendships (and why you need them) www.nytimes.com/2019/11/20/smarter-living/how-to-have-closer-friendships.html (archived at https://perma.cc/EK98-NY2B)

Poulsen, K (2014) No limit: Two Las Vegas gamblers found a king-size bug in video poker. It was the worst thing that could have happened to them, *WIRED*, November, pp 138–45

Premack, D and Woodruff, G (1978) Does the chimpanzee have a theory of mind? *Behavioral and Brain Sciences*, **1**(4), pp 515–26

Psychogios, A, Nyfoudi, M, Theodorakopoulos, N, Szamosi, LT and Prouska, R (2019) Many hands lighter work? Deciphering the relationship between adverse working conditions and organization citizenship behaviours in small and medium-sized enterprises during a severe economic crisis, *British Journal of Management*, **30**(3), pp 519–37

Ramachandran, VS (2011) *The Tell-Tale Brain: A neuroscientist's quest for what makes us human*, WW Norton, New York

Ridley, M (1997) *The Origins of Virtue: Human instincts and the evolution of cooperation*, Penguin, London

Rifkin, J (1995) *The End of Work: The decline of the global labor force and the dawn of the post-market era*, Putnam, New York

Rizzolatti, G and Craighero, L (2004) The mirror-neuron system, *Annual Review of Neuroscience*, **27**(1), pp 169–92

Schulte-Rüther, M, Markowitsch, HJ, Fink, GR and Piefke, M (2007) Mirror neuron and theory of mind mechanisms involved in face-to-face interactions: A functional magnetic resonance imaging approach to empathy, *Journal of Cognitive Neuroscience*, **19**(8), pp 1354–72

Shaffer, M and Schiller, D (2020) In search of the brain's social road maps, *Scientific American*, February, pp 23–27

Stix, G (2014) The 'it' factor, *Scientific American*, Special evolution issue: How we became human, **311**(3), September, pp 72–79

Sullivan, J (2015) Born to trust: The brain evolution as a social organism – a conversation with Louis Cozolino, PhD, *BrainWorld*, **6**(2), Winter, pp 50–53

Thaler, RH (2015) *Misbehaving: The making of behavioral economics*, Allen Lane, London

Thaler, RH and Sunstein, CR (2008) *Nudge: Improving decisions about health, wealth and happiness*, Yale University Press, New Haven, CT

Tomasello, MA (2014) *Natural History of Human Thinking*, Harvard University Press, Cambridge, MA

Tomasello, M and Carpenter, M (2007) Shared intentionality, *Developmental Science*, 10(1), pp 121–25

Van der Goot, MH, Tomasello, M and Liszkowski, U (2014) Differences in the nonverbal requests of great apes and human infants, *Child Development*, 85(2), pp 444–55

Van Overwalle, F (2009) Social cognition and the brain: A meta-analysis, *Human Brain Mapping*, 30, pp 829–58

Watts, DJ and Dodds, PS (2007) Influentials, networks and public opinion formation, *Journal of Consumer Research*, 34(4), pp 441–58

Wimmer, H and Perner, J (1983) Beliefs about beliefs: Representation and constraining function of wrong beliefs in young children's understanding of deception, *Cognition*, 13(1), pp 103–28

Winston, JS, Strange, BA, O'Doherty, J and Dolan, RJ (2002) Automatic and intentional brain responses during evaluation of trustworthiness of faces, *Nature Neuroscience*, 5(3), pp 277–83

Young, LJ (2003) The neural basis of pair bonding in a monogamous species: A model for understanding the biological basis of human behavior, in *Offspring: Human fertility behavior in biodemographic perspective*, ed KW Wachter and RA Bulatao, pp 91–103, The National Academy Press, Washington DC

Zak, P (2011) Trust, morality – and oxytocin? www.ted.com/talks/paul_zak_trust_morality_and_oxytocin (archived at https://perma.cc/888B-664L)

08

Brain communication, better persuasion

'WHY DON'T THEY FOLLOW OUR VALUES?'

The new corporate values were finally ready. Revising and changing the company's vision and values had been at the top of the new CEO's agenda. As soon as she arrived in the company, she gathered her team and swiftly asked them to adopt new values that were more motivational and inspiring than the old ones. She believed that by changing the existing product and company-oriented values to customer and society-centric ones, she would also change the corporate culture. This was one of her key mandates when hired by the board to transform the company from a traditional, slow-moving corporation to a modern, fast-thinking business. She was convinced that values were an excellent place to start!

The corporate communications department worked closely with human resources and even marketing to identify and develop values that reflected the vision of the new CEO. It took them months of internal and external research, analysis and thinking, and with the help of external partners, such as agencies and consultants, they produced a set of modern-sounding, outward-looking, motivational values such as people comes first, customer focus, autonomous work, emphasis on cooperation and synergy, respect for the work of others, etc. The CEO was very happy. She now had corporate values representative of the type of company she wanted to make out of the existing one. Being proud of the new values, she could not stop talking about them in social meetings with peers from other companies. Many congratulated her and this reinforced the feelings of achievement and pride she had for the project.

She immediately ordered the application of the new values throughout the company's materials, online and offline, and she branded offices and other premises with quotations and keywords deriving from the values. Departments had to promote the values at every opportunity by explaining them to employees, highlighting the values' importance for the future of the company. Everyone had to sign a document committing to the values. Team-building events were organized specifically for that reason and the whole project was implemented with success and full agreement and support by employees in all levels. A year later, the majority of people in the company were nowhere near behaving according to the much-publicized values. There was a competitive attitude among employees, less respect for the work of others, no delegation of responsibility and therefore no emphasis on autonomous work. All in all, very little had changed.

What happened? Why did people, while participating actively in their development and agreeing enthusiastically with the final outcome, not behave as expected? What does it really take to persuade employees to collectively adopt new values, or similar initiatives? These are the main questions the CEO asked her team and external partners to answer for her. The company spent a good deal of money and time in building and communicating the values without any significant results. She wanted to know why. The answer surprised her, and changed her attitude towards communications forever. Apparently, communicating with people in the company in a way that induces any type of behavioural change has to be about triggering many brain functions responsible for decision-making and behaviour. Only targeting people's rationality and analytical brain centre doesn't work. Effective influence needs a holistic brain approach and brain-adaptive leaders need to master it. This is a core BAL skill.

Persuading the brain to act

Throughout this book we have discussed extensively how different brain regions and neural pathways affect the way we think, feel, behave and connect. In doing so, we have touched on the importance of communicating in a certain way in order to affect more powerfully our, and other people's, brains. For example, we explained the significance of symbols and words for creating a desired priming environment in Chapter 6, and we referred to feedback in Chapter 1 as a key strategy for saving brain energy and dealing with ego depletion. The common thread in most of these insights is that the human brain is primarily an information-processing organ (Sullivan, 2015). What it does is receive information in the form of stimuli from within the human body as well as from the outside world through the senses and neural networks in our bodies, and reacts to these stimuli accordingly. As we have argued, this mainly depends both on the stimuli itself and the way the brain is calibrated, or adapted, to respond. In our view, the brain as an information-processing centre constantly communicates internally in order to identify the right response to the right stimuli. Communication is a core brain activity, if not *the* core activity. And brain communication, for us, is are all about sending the right message to the right brain structures to persuade the brain to react in a desired way. Since influence is such a central notion of leadership, as rightfully highlighted in the *Harvard Business Review*'s 2013 special issue (Ignatius, 2013), modern leaders are in great need of learning and practising brain-friendly communications in order to effectively persuade others how to, and how not to, behave. Leaders without serious powers of persuasion are like automobiles without wheels: they cannot get very far.

In our quest to identify and implement the most effective brain communications model, for years now we have adopted the triune brain approach, as explained in previous chapters. This approach separates the brain and all its structures into three main areas: the old/reptilian brain; the emotional/affective/limbic system brain; and the rational/thinking/neocortex brain (MacLean, 1990). Each of these structures requires different information/stimuli to get them going and a holistic approach is needed to take into account all three in order to achieve any significant behaviour-targeting communications goals. This is the way in which we apply the triune brain approach and it has proven to be very successful in our experience, despite the criticisms that it has attracted in recent years regarding the distinctiveness of its parts in evolution, meaning that the three parts of the brain did not necessarily develop and evolve in such a clear and distinctive way (Pribram, 2002). Hailed as the single most influential idea in neuroscience since the Second World War (Harrington, 1992), we fully agree with the prominent thinker Gerald A Cory in his assessment that (2002):

despite its current lack of popularity in some quarters of neurobiology, I think that the triune brain concept will continue to be influential, and with appropriate modifications as research progresses, provide an important underpinning for interdisciplinary communication and bridging.

It is the simplicity of the model and its effectiveness in communicating brain functionality to wider audiences, as well as its direct links to cognitive, emotional and behavioural human traits, that makes it a suitable tool to discuss the impact and benefits of neuroscience in leadership and organizations. Montag and Panksepp (2017) agree, calling the three brain concept a simplification, but a didactic one, meaning that its educational value as an introductory way of looking at brain structure and function is evident.

THINKING EQUALS PERSUADING

In Chapter 7 we explained what might prove to be the most surprising plot twist in the modern history of humankind: consciousness is about 'we' and not 'I'. Conscious reflection forms the basis for the functional inhibit-and-plan processes that help us navigate better both the physical and the social environment. Concerning the latter, consciousness helps us decode what other people might be thinking, feeling and intending to do as key inputs for dealing with them effectively and efficiently. But this is not the whole story.

In their highly influential argumentative theory of reasoning, Hugo Mercier and Dan Sperber view reasoning as the foundation for logical analytics, conscious deliberations and conversational interactions. In their approach, reasoning is an intuitive brain process for winning arguments, influencing others and facilitating common action (Mercier and Sperber, 2017). This means that our elaborate ability to formulate conscious thoughts and derive arguments from them in order to convince ourselves and others to initiate desired behaviours is rooted in unconscious processes that have evolved to promote collaboration and shared intentionality. These unconscious processes become partly conscious in the form of intuitive reasoning to help us create teams with common goals and perform complex tasks together. In their own words:

> Whereas reason is commonly viewed as a superior means to think better on one's own, we argue that it is meanly used in our interactions with others. We produce reasons in order to justify our thoughts and actions to others and to produce arguments to convince others to think and act as we suggest... The main role of logic in reasoning, we suggest, may well be a rhetorical one: logic helps simplify and schematize intuitive arguments, highlighting and often exaggerating their force.

The main differences between humans and other animals in cooperation are that humans collaborate with complete strangers, not just with family and close tribe members, and also collaborate in long-term projects, not just in short-term survival pursuits. Additionally, the amount of information exchanged between humans is remarkably richer than any other animal, as is our dependency on this information to learn, grow and become successful in our social domains. Argumentation, based on reasoning and logic, aiming at persuading ourselves and others, is what conscious thinking is designed to do within the wider framework of inhibit-and-plan discussed throughout this book. Simply put, our inner deliberation of how to move forward in implementing a decision, which has already been initiated unconsciously, requires us to take others on our side, because nothing achieved by humans can be done alone. Thinking, as in reasoning, equals to persuading.

Leaders need to target all three brains in order to persuade and change behaviours. In interpersonal and organizational settings this is successfully captured in Chip and Dan Heath's (2010) model presented in their bestselling book *Switch: How to change things when change is hard*. Without explicitly referring to the triune brain, the Heath brothers expanded the concept of 'the rider and the elephant', which was first introduced by psychologist Jonathan Haidt (2006). Heidt metaphorically described how our rationally thinking brain functions as a rider – clever and fast but not able to move alone – and our emotional brain functions as an elephant – mobile and powerful, but irrational. Each system needs the other: they have to reinforce one another in order to achieve any meaningful result. These two systems were also described in detail by Nobel Laureate in Economics Daniel Kahneman, in his internationally acclaimed 2011 book *Thinking, Fast and Slow*. The Heath brothers added a further element, called the *path*, highlighting the fact that, besides thinking and feeling, the environment itself can shape our actions – something very close to our discussion of brain automaticity in Chapter 6. The Heaths, based on extensive behavioural experiments as well as on similar arguments and examples developed by other authors and scholars, show that changing behaviours needs to follow the recipe below, adapted to the themes of the BAL approach in this book:

1 *Rationalize the direction (or direct the rider, for the Heaths)*. The rational brain too often gets stuck in meaningless analysis and endless thought-satisfying debates. It cannot make up its mind easily by itself. To avoid analysis paralysis, we need clear and simple directions for the whole system to follow. Communicating to this brain is all about clarifying

blind spots, utilizing 'bright spots' or success stories, and using simple, direct language. Avoiding information overload and showing the ultimate destination are critical moves in engaging the rider in the most effective way. In essence, we need to convince the rational, conscious mind not to inhibit the behaviour we are trying to initiate (or to inhibit the behaviours we are trying to eliminate), and to use its planning properties to adjust behaviours accordingly.

2 *Emotionalize motivation (or motivate the elephant).* The emotional brain is the one that moves the system forward, not the rider. However, an elephant without destination and purpose is as good as a lost elephant. So, it needs the rider for implementation instructions. However, instructions alone cannot move the elephant. Instructions are not decisions. We move people when we help them adopt a growth mindset, when we break down change into less fear-inducing chunks, and especially when we find the right emotion or mix of emotions suitable for specific teams and individuals. The realization that emotions move people more than ideas, numbers, orders and procedures is crucial for leaders and managers in adopting the BAL model. We have touched upon this extensively in the book.

3 *Formulate the environment (or shape the path).* The environment consists of all those aspects that can influence our behaviour and/or our perception. In other words, sometimes the behaviour of someone is the result of neither his/her rational, thinking process nor his/her motivation rooted in an emotion. The behaviour is rooted in other external factors that may also have an impact on human behaviour. For example, the physical setting of a manager's office could potentially influence the perception and therefore the behaviour of employees while being in that office. Consequently, our brain automations have to get the right cues from the environment in order to allow the elephant and the rider to move towards the desired destination. Brain automations can either be obstacles or enablers in this path. Habits and the physical environment of the organization as well as the social environment can either hinder or boost our teams' efforts in achieving the behavioural goals we have set.

Unfortunately, most of the corporate initiatives that we have come across around the world pay too much attention to the first element – the rider or the rational brain – and overlook the second and the third. Even when it comes to trying to influence the rider alone, most of the time organizations do exactly the opposite to what modern neuroscience, psychology and behavioural economics all tell us: they overload the rider with, often unnecessary, information and/or they exclude critical information needed to set clearly the destination. For example, in a meeting, managers repeatedly overemphasize the details, or give extensive information regarding a project or task, attempting to help employees achieve a result by analysing the situation around the task. But doing so, they often miss focusing on key points that are critical for the task's completion, misleading and overloading the employees' riders and most probably increasing their stress levels. Thus, we do not seriously involve the other two parts of the brain, and we do not address the rider in the right way. This is why, as in the opening case of the chapter, even if we do whatever we can to persuade our employees to follow new corporate values, people won't necessarily do so. The rider, the elephant and the path must be holistically dealt with. Successful behavioural influence requires all of them in one way or another. The faster you adopt this holistic mentality, the better for your leadership persuasion skills.

MEASURING AND CHANGING THE COLLECTIVE CORPORATE BRAIN

The collective brain of any organization, represented by the combined states of the three brains of the employees, is directly related to the culture of the organization. The way people think, feel and behave within their company-specific environment directly affects the success of a corporate culture. Wouldn't be amazing if there were a way to measure this collective brain in order to better understand its current state and to change it if and when needed? This is exactly what we have done in the South East European (SEE) region since 2011.

One of us (Dr Dimitriadis), in close consultation with the other (Dr Psychogios), has developed a diagnostic tool for measuring and changing the collective corporate brain through internal communications and human resources initiatives. The tool was implemented successfully in multinational and regional companies from many industries, such as retail, distribution, banking and food production in the SEE region, and won a prestigious industry award in that region (Dimitriadis, 2014). The tool captures effectively:

1 *What people know.* This part refers to the cognitive brain of the company. It measures the level to which employees are happy with the information they receive concerning both the wider position and future of the company and their own role in, and contribution to, the company (the big and specific pictures, respectively).

2   *What people feel.* This part refers to the emotional/affective brain of the company. It measures the level of a large number of feelings within the organization. We measure all core emotions as well as subjective feelings such as appreciation, support and fatigue.

3   *What people do.* This part refers to the habitual/automatic brain of the company. It measures the level of existence of behavioural habits of employees in their everyday actions such as conflict resolution, intra- and inter-departmental communication, internal and external brand advocacy and others.

The tool, based on employee questionnaires and management interviews, was developed as a response to what we were experiencing in the marketplace: a mismatch between problems and solutions. Companies that had emotional issues attacked these with more information, while others that lacked well-informed employees responded with emotional statements and visionary speeches. We cannot solve emotional problems with more information, just as we cannot solve informational gaps with emotions. Without a clear compass of where the problems exist in the collective brain, we are blind and bound to waste both money and time. On the contrary, by examining the three brains of our company we can establish which one has the actual problem and work on targeted solutions.

The tool has been used to redirect internal communication efforts, to unite various functions on finding real solutions depending on the problem, and to give the companies' leadership a clear direction for building a winning corporate culture. Dividing brains, business units and departments allows us to identify within the organization specific pockets of concerns and pockets of brain excellence, and use them both appropriately. Maybe one department needs its confidence boosting (emotional treatment) while the other needs new habits of reporting problems (behavioural and procedural treatment). One-size-fits-all does not exist for individual brains, and as such it does not exist for collective corporate brains either.

ACTION BOX: CREATE YOUR OWN COLLECTIVE BRAIN DIAGNOSIS

You can create your own version of the corporate brain diagnostic tool for your company, department or team. This is how.

1 Start by creating three categories: *think*, *feel* and *do*. First, how my people think, what they know and what is on their mind. Second, what my people feel or which are the predominant emotions and which are least present. Third, what habits we have, how we usually behave to one another or what are the accepted norms of corporate behaviour within and outside the organization.

2 Under each category, use the information in this and in other chapters of the book to create a list of variables (or questions) that are relevant to your industry, company, situation. Do not overdo it at the beginning. Ten variables (questions) per category are enough to get you started. Those variables could include:

   a For 'think': How aware are we of the strategic intentions of our organization, do we know of the big issues that affect the company, is it clear how our job contributes to the goals of the company, do we know all the requirements and responsibilities of our job, are channels of communication fully understood?

   b For 'feel': How supportive is the company of its people, is optimism or pessimism the dominant mood, do you feel energized/tired, noticed/ ignored, curious/bored, part of a team/isolated?

   c For 'do': Are we proactive or reactive in problem solving, do we approach others when we are faced with an operational problem that we are responsible for, do we collaborate with other departments effectively, do we recommend the company to outsiders as an excellent place to work ?

3 Interviews (or targeted conversations) are the recommended technique for the first time you implement the tool. Discuss, based on the variables, with at least 10 people from the same department. Note down the basic information for each variable and score each one as high, medium or low depending on their performance. We recommend that organization-wide, quantitative studies should be done with the involvement of experts in such research.

4 The outcome is to be able to create first, an overall assessment for the three brains (like thinking brain: high; emotional brain: low; habitual brain: medium) but also a more detailed picture of which variables/elements in each category are high, medium or low. The most problematic areas need your immediate attention. Use the rider, elephant and path methodology to change your people in the areas they scored medium or low. Try to sustain and use as examples the variables that scored high.

Expand the tool to different teams and departments. Then compare and contrast them. Do you see any differences? Are there departmental-specific patterns favouring or limiting any of the variables in the three brains of the company? Act accordingly.

Cialdini's influence

If you research how to improve your influencing skills this will inevitably lead you to Robert B Cialdini, the undisputed global guru of persuasion. Cialdini, the Regents' Professor Emeritus of Psychology and Marketing at Arizona State University, published his seminal book *Influence: The psychology of persuasion* in 1984, and since then his six principles of persuasion have become the industry standard. Although he developed these principles mainly for marketing and sales use, they can be easily adopted and applied by leaders within their organization. The principles (Cialdini, 2007) we have been using for years, explained in the context of leadership and management, are as follows.

- *Reciprocity.* Reciprocating an action of goodwill is hardwired in our social brains, as we explained in the previous chapter. Our deep-rooted sense of fairness and moral behaviour asks us to return a good gesture to the people that first behaved kindly to us. Applying Axelrod's recommendation for Chapter 7, a leader needs to start a relationship with goodwill. This can help to open up a positive and potentially perpetual circle of cooperation and at the same time, according to Cialdini, it increases our power of persuasion. The gravitational pull of good actions is so strong that it can mould other people's behaviour according to ours. When behaving badly, you can also expect to influence the behaviour of others, this time negatively. In other words, if you personalize, others will too; if you smile, others will too; and if you innovate, others will innovate too.

- *Commitment and consistency.* Much of human behaviour can be attributed to our efforts to reduce what psychologists call cognitive dissonance. This is an important concept, first developed by American social psychologist Leon Festinger in the 1950s, and it suggests that people strive in their lives to achieve internal consistency (Festinger,

1957). This means that we constantly try to match our beliefs with our actions or our many beliefs together. Inconsistency (dissonance) between what we believe and what we did on a certain occasion can cause serious internal conflicts that have to be somehow resolved: either by changing our belief, trying to ignore the problem altogether or behaving consistently soon after. For example, committing your team to a certain course of action will influence them to behave accordingly. In other words, if you want your team to show commitment in a project, you need first to show commitment. But be careful: commitment has to be considered deeply important to them. If not, the internal conflict of not following with their actions will not be that great. Make them believe in the action (being integral to their purpose, for example) and they will do everything they can to act consistently. Also, you can ask your team to write down the reasons why they will commit to an initiative and to present those to the team. Reasons like responsibility towards the rest of the team as well as towards end customers that the project targets, reputation of the team and the company, commitment to our values, etc, could work nicely here and become collective reasons.

- *Social proof.* We are all aware of the effect that a group of people looking up in the sky has on a bystander: the bystander will intuitively also look up. This is social proof, or the principle of people following the social group unconsciously and automatically. The classic and famous Asch conformity experiments demonstrate that people can follow the opinion and behaviour of other people even when these other people are obviously wrong. In the original experiment, 50 participants had to separately go through a series of tests where they had to choose between various options and point to the correct answer. The problem was that in most of these tests, other people present gave obvious wrong replies openly in front of the group. Would the participants follow them or would they state the obvious correct answer (for example: Look at these three vertical lines on the paper. Which one is the shortest?). Surprisingly enough, only 25 per cent of the participants were not swayed by the others' incorrect answers, while 75 per cent stated the wrong option (the answer that had been supported by others) at least once (Asch, 1951). We have conducted a similar experiment with our students many times in the last few years. We ask all students in a class to read a demanding text on politics or economics. Then we ask them to express their opinion in the class about the main issues in that text. However, the first couple of students to express their opinion have been prepared in advance to mislead the others by expressing completely and obviously wrong assessments. Then we ask the other students (the real subjects of the experiment) to express their own views. Guess what happens – the majority of the students, although tending to look somewhat surprised, are influenced by the first two misleading responses and agree with them. Only a very few disagree. What seems to be critical here is the timing of the first correct answer in the class. If, for example, it appears straight after the first two misleading ones, the chances of the rest being correct are higher. If the correct one appears later then it tends to have less influence, since more opinions agreeing with the first two misleading ones have emerged prior to the correct one. It is highly intriguing but also quite revealing about our social behaviour, isn't it?

We have social brains, and thus we have a tendency to follow the crowd even when it is obviously mistaken. In another favourite experiment of ours, the rate of stealing pieces from a Petrified Forest park in Arizona increased after a counter-theft campaign stated: 'Many past visitors have removed the petrified wood from the park, changing the state of the Petrified Forest' (Cialdini *et al*, 2006). A campaign to decrease theft actually promoted it! So much for common sense messaging. Our brains tend to follow by default what other people do. When announcing an important meeting, always mention the people that have already confirmed participation, especially if the figure is high. If participation proves to be limited, do not shout it out. Negative social proof has a strong influential pool, meaning that some people will probably opt out of the next meeting by default if their brains capture the fact that past meetings were poorly attended. Cultural norms within organizations tend to replicate people's accepted behaviour, positive or negative. Pay extra care to which one is which and change it if necessary.

- *Authority.* People tend to follow rules by default when those rules are made by figures they deem highly authoritative. The seminal experiments on obedience by Stanley Milgram and Philip Zimbardo mentioned in Chapter 5 prove the point to the extreme. This means that it has to be clear in the company who is responsible for which decisions. Confusion of authority harms your persuasion potential. Furthermore, some classic studies (French and Raven, 1959; Mechanic, 1962; Hickson *et al*, 1971; Pettigrew, 1972; Patchen, 1974) show that expert power can boost the authority effect and increase persuasion. A communicator that is perceived as possessing high expertise can influence both attitudes and the memory of others. By stimulating the caudate nuclei in the brain, an area in the basal ganglia related to feedback processing, learning, expectations of reward, social cooperation and trustworthiness, as well as the medial temporal lobe associated with long-term memory formation and high expertise, the positive acceptance of a message, meaning attitude, increased up to 12 per cent, and its recognition, meaning memory, increased 10 per cent (Klucharev *et al*, 2008). These are actually very high numbers compared with older studies (Klucharev *et al*, 2008). Be an expert in your field, whatever field this is, meaning strive for excellence in your job area rather than just being good at it. Demonstrate this expertise whenever needed, especially in high-pressure situations, and your influence levels will increase.

- *Liking.* We are more likely to be influenced by the people we like and admire than by those we don't. The three principles of liking and their impact on persuasion are well evangelized (Dobelli, 2013). First, we like people that are physically attractive. Second, we like people who share some similarities with us. And third, we like people who like us. This means that leaders and managers have to take care of their physical presence and always be presentable. They also need to present themselves in ways that imply some similarity to their followers. The leader can demonstrate this through statements like, 'I started the same as you', 'We share the same vision', and so on. Lastly, leaders should give compliments whenever possible and appropriate. Part of this persuasion principle includes the claim that we make more effort for people who like us than for people who do not like us. In our experience, this has been consistently proved right. We have observed that people are at their most passionate, proactive and inventive when they are being appreciated.

- *Scarcity.* Our brains have evolved to notice scarcity and to automatically move towards securing vital resources that are about to be in short supply. Prioritizing behaviour that gains access to resources that our perception categorizes as scarce was a necessary evolutionary strategy, since failing to acquire those resources could mean failing to survive. Cialdini emphasized the marketing and sales aspect of this principle (as with most of the principles which were originally conceived by him for marketing use), advising marketers to always remind potential customers that the fantastic offer they have is expiring soon or that this product is selling fast and is never to be produced again. In modern organizational environments, leaders should always remind their teams that resources, their current position in the market, even the very existence of the company are not to be taken for granted. Business magazines are filled with cases of once mighty companies who, in the space of only a few years, had to struggle for their survival (like Nokia, Kodak and others). It is not so much about scaring people. It's all about breaking comfort zones, doing reality checks often and setting the tone for dynamic, proactive and innovative work within the team. Furthermore, creative competitions with short timescales that offer desirable prizes to employees can be used to create short boosts of enthusiasm and new ideas. 'Hackathons', as they are called, or brief and intense project development competitions originating in internet, software, digital and creative companies (Leckart, 2012), work so well because they happen rarely. We believe that if they took place every day they would not have the same effect, plus they would certainly bring the workforce to extreme exhaustion. So, using hackathons, in a relevant form for your company, is highly advisable but as a scarce, meaning not every day, solution.

In 2016 Cialdini updated his classic model by adding one more persuasion principle, which he called unity.

- *Unity.* This principle concerns the effect on persuasion of social proximity and shared identity. It suggests that the closer the affiliation of a person to a social group or entity the higher the persuasive impact of a message that encompasses this closeness. Shared identity, according to Cialdini, is formed based on being together and acting together. Being together is about sharing common characteristics strong enough to categorize us in the same social group. As soon as we identify strongly with the group, we will be easier and more quickly persuaded to act in accordance with the group's interests and behavioural directions. Acting together refers to the anthropological phenomenon of people bonding with each other when they perform coordinated actions in unison. Dancing together or playing music together have been historically proven to create synchronicity and to strengthen shared identity, but also engaging in social interactions such as speed dating, in a highly structured and reciprocal way, strengthens emotional bonding surprisingly fast even between complete strangers. In a working environment, leaders need to exercise both ways of increasing the sense of unity within teams, between departments and even with external partners. Being together can involve shared values, common purpose and a narration that binds people together, while acting together can involve tightly coordinated actions that involve the whole team in a limited timeframe to produce extraordinary results. Additionally, team-building efforts should include activities to strengthen both the being together and the acting together dimensions of unity.

The seven principles of persuasion cannot easily work in all situations and with all people. But we have found that with practice and a bit of reflection you can become skilled in choosing the right principle for the right challenge. Some of them can even become habits of your new BAL leadership approach. Use them and increase your persuasive power.

Talk to the brain

Whenever we discuss the topic of leadership persuasion and the brain with our students and corporate clients we often get the question: 'OK great! But do you have any fast tricks or specific wording that we can apply immediately?' Although we are not supporters of those 'revealing' magic words that create blind followers, there are scientific findings that support the use of specific words or phrases that have immediate influential effects on people. Furthermore, there are conversational styles that can help us increase our persuasion impact and collaborative potential in the long run.

Let's start with those that have a quick-win effect and that we personally use almost daily in our teaching and consulting practices. Psychologist, author and media commentator Rob Yeung did an excellent job in summarizing those fast-track techniques at the end of his book *I is for Influence: The new science of persuasion* (2011). Here they are:

- *'That's not all.'* This makes a good offer even better. As Yeung explains, proposing a new position to someone, presenting an organizational restructuring project, or trying to convince your team to put in double hours this month to boost the numbers can be done much more effectively if you explain the advantages – then follow them up with even more. Do not play all your cards in your main speech but save surprising benefits for the very end. Using the words 'and that's not all' in the right way has the power to make people feel that they get more than they expected, more than they should or more than they negotiated for.

- *'Because…'* A very intriguing and widely cited experiment showed that if a person is trying to jump the queue for the photocopier, they can obtain a high approval rate if they specify their reason. 'Excuse me, I have five pages. May I use the Xerox machine, because I am in a rush?' won a staggering 94 per cent acceptance rate than the simpler 'Excuse me, I have five pages. May I use the Xerox machine?', which got 60 per cent (Langer et al, 1978). Adding the word 'because' boosts influence drastically. The more serious the situation, the more serious the 'because' has to be. It is a shame how many times managers and leaders just say 'This is not possible', 'Not now', 'I will think about it later' or other phrases that create disappointment for their colleagues and subordinates. Just adding 'because' and offering a true and honest reason would leave everyone happier. When asking your team for favours at work or for extra effort, always explain to them clearly and directly why this is needed. You should incorporate this technique into your communication because it will make a huge difference to your effectiveness as an influencer.

- '*I need you to [do something very specific and/or quantifiable].*' Being specific in a request vastly increases the positive impact of that request. Instead of saying 'We have to increase our efforts', better say 'We have to increase our productivity rate by 7 per cent in comparison to last year in order to catch up with other departments.' Being specific talks effectively to the rider in your brain and sets aside any misunderstandings, confusion or unnecessary pessimism. If the goal is clear and in small chunks, it is also manageable.

- '*You have a choice.*' Always reminding colleagues, employees and business partners that they actually have a choice in front of important and pressing decisions considerably increases the likelihood of the acceptance of the option you are aiming for. This allows people to feel that they are in control, rather than feeling trapped. A feeling of being trapped or in danger can over-stimulate the amygdala with all the negative consequences previously mentioned (such as stress, fear, anxiety, aggression, etc). We always have a choice in any decision, even in the most difficult ones. Explicitly saying so, particularly with a supporting tone of voice, persuades people that what they will opt for is their own decision. This boosts confidence, engagement and satisfaction. Instead of saying something like 'We have no choice, we need to restructure', it would be better to say 'We have a choice: either we stay as we are and we most probably become irrelevant to our customers within six months or we change now and we beat the competitors at their own game!' Can you spot the difference?

Fast persuasion is very handy in our everyday meetings, discussions, announcements and presentations. However, we do not advise becoming a trickster of persuasion. We can rarely completely hide our true capabilities, intentions and leadership potential. In this respect, adopting a conversational style that allows us to become more deeply receptive of others and to respond in a more meaningful and influential way can improve the very core of our brain persuasion capability. This is needed from leaders especially because, as Andrew Newberg and Mark R Waldman, authors of the book *Words Can Change Your Brain: 12 conversational strategies to build trust, resolve conflict and increase intimacy* (2012), explain:

> Although we are born with the gift of language, research shows that we are surprisingly unskilled when it comes to communicating with others. We often choose our words without thought, oblivious of the emotional effects they can have on others. We talk more than we need to.

Newberg and Waldman suggest a solution that they call *compassionate communication*, which, when practiced long-term, aligns the two sides of our brain and increases neural resonance between conversing people. Neural reasoning refers to the 'accurate transference of information from one brain to another' (Newberg and Waldman, 2012), boosting the possibility of the right signals being sent and received, and of cooperation emerging. The authors suggest 12 ways by which we can become compassionate communicators, many of them touched upon earlier in this book. These are: adopting a more relaxed attitude, being fully present during a conversation, cultivating inner silence, increasing our positive mentality, reflecting on our deepest values (or purpose), accessing pleasant memories (in order for positive facial expressions to emerge), observing nonverbal cues in others, expressing appreciation, speaking warmly, slowly and briefly, and listening deeply. This list is a very convenient summary for becoming a better, more cooperative, and essentially a more influential converser.

We believe in those and in other similar attributes discussed in this book. Although implementing them all is always easier said than done, we have seen real improvements in ourselves and in others whenever we work closely and seriously on this matter. This starts with the simple but profound realization that communication is hard and most often ineffective if we do not consider how the brain works whenever we aim to have any impact on other people's behaviour. In a very well-received TEDx talk at the University of Strathclyde in Scotland, titled 'The illusion of communication and its brain-based solution', one of us (Dimitriadis, 2015) urged the audience to adopt brain-friendly communications if we are to change this world for the better. Sadly, what we see every day within companies is riders speaking to riders, behaving like there are no other brain parts to be included in decision-making, and especially in adopting agreed courses of action. So, when two people are conversing, two frontal lobes agree with each other on what is logical, rational and analytically correct to do, and then two whole brains (not only frontal lobes, which actually become irrelevant after the conversation) just do not move those same two people to act accordingly. Picture this: talking to your colleagues you collectively and swiftly agree that putting 110 per cent of your combined effort behind the new project is crucial for the success of that project. But then, it becomes clear that people have not increased their efforts as much as was originally agreed upon. Sure, few of them tried a bit harder, but nothing close to what was thought to have been agreed. The rider understood but the elephant did not move and the path was not clear. Only when we start considering the brain as a whole, both ours and those of the people we talk to, can we make meaningful decisions that will ultimately lead to behavioural change. Modern leaders need to talk to brains as a whole and not exclusively to the rationality-dominated frontal lobes. Talk to the brain and people will move.

THE ALL-IMPORTANT FEEDBACK

In recent research work of ours with colleagues (Psychogios *et al*, 2018, 2019), based on many interviews with managers from different type and size of organizations as well as from different industries and sectors, it was found that managerial feedback is a core cue through which routines at the workplace can change. Feedback provides information about work characteristics and attempts to steer performance in a given direction (Fedor *et al*, 2001), and is seen as an integral part of the learning process (Ashby and O'Brien 2007). In our study it was found that feedback could be used as a way to influence the change of organizational behavioural patterns. In particular, we found that there are specific dynamics that occur during the process of feedback and help managers to make sense, facilitate, endorse and monitor this process (Psychogios *et al*, 2018). Moreover, in a more recent study, we found that three principles that can be applied when giving feedback were crucial (Psychogios *et al*, 2019):

- First, feedback works better when it is informal. When you want to talk to an employee regarding behaviour that they repeat which is creating problems, then it is preferable (at least the first time around) to avoid formalized procedures like asking them to come into your office. Opt for more informal ways. The study showed that managers could achieve better results through informal structures like talking over a coffee, or walking and talking.

- The second principle is that feedback should be as specific as possible. Managers need to directly target, from the beginning of the discussion, the behaviour that needs to be changed. Long and generic discussions should be avoided and it needs to be made clear what the issue of the discussion is and that a decision should be made by the end of it.

- The third principle is that the feedback should also demonstrate the benefits that will be achieved after the change of the behaviour. Managers need to discuss these benefits with employees, emphasizing what the latter can gain after changing a habitual behaviour that is currently causing harm.

We call the above three principles the threefold framework of feedback in organizations.

It is easy to spot similarities between this study's results and many of the concepts discussed in this and previous chapters. This is because feedback is a basic function influencing our brains and subsequently our behaviour. It has been found that carefully utilizing informational feedback loops can allow the brain to react appropriately to the initial message and to change our behaviour. Effective feedback loops have four components (Goetz, 2011):

1 *Evidence*. Data regarding behaviour must be immediately received and presented.

2 *Relevance*. Cold data are turned into meaningful input by way of design, presentation, comparison and context.

3 *Consequence*. A message needs to carry a purpose or wider goal. This answers the question 'Why?'

4 *Action*. The individual acts in order to achieve the desired data, thus closing this loop and opening a new one.

As an example, think of a company's effort to improve sustainability related behaviours at the workplace. If there are recycling targets, the recycling output could be communicated often and with concrete data (evidence), data could always be presented next to the desired target for immediate comparison (relevance), a motivational comment can accompany the data announcement to remind people of the purpose and importance of the activity (consequence), and the loop closes with new actions that hopefully will lead to improved measurements (a new loop is born).

Persuasion stimuli

The recently discovered persuasion powers of brain communication have led to an explosion of *neuroculture*, or the wide diffusion of neuro-scientific insights into society as a whole and into popular culture (Frazzetto and Anker, 2009). We find this process very exciting since it is obvious that we believe that leadership, business, education, politics and life in general will drastically improve if a more brain-based approach is adopted. Effectiveness of communications can increase exponentially, leading to substantial savings both in money and time. This is more so in advertising where as the old adage goes: 'Half my advertising is wasted, I just don't know which half' (AdAge, 1999). Neuromarketing – the marketing, advertising and sales sector's effort to develop new, more effective communications based on the neuroscience of persuasion – includes many tips, techniques and models of how companies can achieve more with less. Some of them can be easily applied in a leadership context, especially the six stimuli model of Patrick Renvoise and Christophe Morin, first published in 2002. This model gives extremely helpful insights into how leaders should communicate directly to the brains of their associates and employees. In Table 8.1 you can see the pioneering work of Renvoise and Morin (2007) on which stimuli to apply in order to be drastically more effective and efficient in persuasion, in the context of internal communications and management information.

A case in which you would be able to use all these stimuli at the same time would be rare. However, you can use a mix of two or three of them in order to make sure that your message will reach the parts of your audience's brains that will induce behavioural responses. Otherwise you will just talk to the executive part of the brain that might be doing a lot of analysis but not much action.

TABLE 8.1   The Renvoise–Morin six stimuli model adapted for leadership

Stimuli	Description	Leadership Lesson
Self-centredness	Older and deeper brain structures are primarily concerned with our survival. The individual sense of self-preservation makes our brain notice what refers to us personally.	In meetings, presentations and announcements of important news we always need to consider emphasizing the 'what's in it for me' aspect of our message and to make sure we are taking the receiver's perspective. Use the word 'You' frequently and in a positive and constructive way. And of course there are individual aspects in teamwork, although some argue there should not be. But trying to ignore the individual self within teams and suppress it can be disastrous. Trying to promote it through mutual understanding and reciprocal collaboration can elevate it.
Contrast	Our brains have to spend energy in order to spot differences between messages. Thus, they are inclined not to, in order to save brain power.	Stop sounding the same in every internal communications campaign and management information delivery. Make it easy for the brains of your employees to see what you are saying by always developing messages that are clearly emphasizing distinct features and distinct states. The 'before-and-after' and 'with-and-without' communicational effects apply here and can be used to convince people about new policies and change projects. 'Look at the situation without this new rule and with it. Which one do you prefer?' Such messages utilize contrast effectively for making information easier digested by the brain.
Tangibility	The brain is better at understanding concrete and tangible information than abstract and theoretical information. It spends less energy in capturing the former and much more in dealing with the latter.	Always try to use concrete and clear pictures and words, as well as specific contextual information that make a difficult and abstract topic precise and tangible. Numbers can help, especially if mentioned in a specific context: from 'We will spend a lot on this project' to 'We will spend one million, which is our entire departmental budget, on this project.' The message has a much stronger impact on the second version.

TABLE 8.1   *continued*

Stimuli	Description	Leadership Lesson
Beginning and End	One of the reasons for storytelling being such an ancient art (think of the *Iliad* and the *Odyssey*) is because our brains like to receive information sequentially: with a beginning and an end. So attention is naturally higher at those two points.	Be very careful in choosing your words at the beginning and the end of important announcements. The same goes for when meeting new colleagues and employees: they will clearly remember your first and last words in the meeting if they are powerful. Never give a dull start to a presentation in the hope that later on you will give the juicy details. Attention will be lost long before you reach slide number 18.
Visualization	We are visual beings, and the brain receives most of its information, and more quickly, through our eyes than other senses. The more graphically visual the message, the better.	Never use blunt, boring or difficult to read or understand materials and slides. The message has to be clear, big and centre-stage. Use fewer but bolder visuals, fewer but bolder words and numbers. Leave minimalistic art for the art exhibitions and overburdened Excel slides for MBA students. Following the cultural norms and brand guidelines of your company, create internal campaigns and presentations that are direct and easy to notice and understand. The more striking and clear the visuals, the better the effect. Again, be careful of your organization's unique acceptance reflexes for such efforts.
Emotionality	Brains are emotional organs in the sense that it is emotions that make us move to respond behaviourally to a message. As we saw earlier in this chapter, the elephant moves us and the rider directs us.	It is not true that corporate information and communication cannot have emotions. If they couldn't then we would live in an unhealthy, even psychopathic, environment (as explained in Chapter 4). Emotions are an integral, if not central, function of our brains and thus using them correctly can only lead to more impactful communication and persuasion. We have noticed that many managers try to de-emotionalize their messages and thus dangerously lower the impact of what they have to say. By being sensitive to the culture of our organization we can infuse our messages with emotions relevant to what we are trying to achieve in order to maximize the effect. Emotions equal actions. It is as simple as that.

SOCIAL INFLUENCE DIMENSIONS

Influence and persuasion are not always clear-cut practices with black-and-white options. They require a nuanced approach tailored to the individual and the team we address. They also depend heavily on situational factors. In this direction, Jonah Berger, Professor at the Wharton School of the University of Pennsylvania, in his book *Invisible Influence: The hidden forces that shape behavior* (2016), discusses three dimensions that should shape our influence approach and can impact significantly on the success of our persuasion efforts. These are:

- *Similarity vs differentiation*. We have already discussed the importance of shared values, shared identities and shared intentionality. However, people are also influenced by their need to be different. Apple's 1997 campaign 'Think different' is based on the human drive to not always conform to the masses and desire to feel different. The balance between in-group imitation and out-group differentiation is crucial for maximizing our influence on the targeted group. Sometimes, we can influence behaviour more if we remind people who they should not be like, rather than which group they actually belong to.

- *Familiarity vs novelty*. The power of familiarity lies in the fact that our brain processes more efficiently, in terms of time and energy, stimuli that are similar to older ones than those that are completely unknown and unfamiliar. Uncertainty is lower in such cases and previous experiences can be used as roadmaps for quick decision-making. On the other hand, novelty has its own advantages since it provides the necessary variety to improve one's chances of learning and growth, it attracts faster attention because of uncertainty and it is proven to increase life satisfaction when present at intervals. Leaders need to balance familiarity and novelty in their influence efforts by observing their team's past behaviour when exposed to known or unknown information.

- *Collaboration vs competition*. Being and acting together in unison are powerful persuasive conditions. At the same time, increasing the sense of competition between specific individuals and/or groups can boost our persuasive impact and move people faster and more strongly. Integrating collaboration and competition is a crucial aspect of a leader's influence toolbox.

Berger's viewpoint is a nuanced one: do not use only one approach or the other. Use both alternatives in all three categories, even simultaneously, to maximize influence. Knowing when to do so separates persuasive leaders from reactive ones.

Keep in mind

Persuasion is a core leadership skill and modern leaders need to be able to influence behaviours in order to achieve their organization's goals. This means we need to 'speak' to all three main functions – thinking, feeling and behaving – in order to achieve important results. Frontal lobes, or executive parts, speaking to and agreeing with each other can rarely lead to passionate and meaningful behavioural responses by themselves. Emotions have to motivate and the environment facilitate the desired course of action. Direct the rider (rationality), motivate the elephant (emotions) and tweak the path (habits and procedures) to achieve maximum results both for individual and collective brains. Use when possible: Cialdini's seven principles of persuasion; specific phrases that can fast-track your influence on others (like 'because'); the 12 points for compassionate conversations; and the six stimuli model of getting a brain's attention. Also, the outcome should be based on a combination of approaches rather than one or another. Your increased power of persuasion, if used appropriately, will have a significant positive impact both on your and other people's work.

References

AdAge (1999) John Wanamaker – special report: The advertising century, https://adage.com/article/special-report-the-advertising-century/john-wanamaker/140185 (archived at https://perma.cc/2C8S-H28B)

Asch, SE (1951) Effects of group pressure on the modification and distortion of judgments, in *Groups, Leadership and Men*, ed H Guetzkow, pp 177–90, Carnegie Press, Pittsburgh

Ashby, FG and O'Brien, JRB (2007) The effects of positive versus negative feedback on information–integration category learning, *Perception and Psychophysics*, 69, pp 865–78

Berger, J (2016) *Invisible Influence: The hidden forces that shape behavior*, Simon and Schuster, New York

Cialdini, RB (2007) *Influence: The psychology of persuasion*, HarperCollins, New York

Cialdini, R (2016) *Pre-Suasion: A revolutionary way to influence and persuade*, Random House Books, London

Cialdini, RB, Demaine, LJ, Sagarin, BJ, Barrett, DW, Rhoads, K and Winter PL (2006) Managing social norms for persuasive impact, *Social Influence*, 1(1), pp 3–15

Cory, GA (2002) Reappraising MacLean's triune brain concept, in *The Evolutionary Neuroethology of Paul MacLean: Convergences and frontiers*, ed GA Cory and R Gardner, pp 9–27, Greenwood Publishing Group, Westport

Dimitriadis, N (2014) Neuromarketing is the future, www.ekapija.com/website/en/page/842820/Nikolaos-Dimitriadis-CEO-of-DNA-communications-Neuro-marketing-is-the-future (archived at https://perma.cc/8DWY-TF4D)

Dimitriadis, N (2015) The illusion of communication and its brain-based solution, TEDx, The University of Strathclyde, Glasgow, 25 April

Dimitriadis, N and Psychogios, A (2020) Social brain-constructed relational leadership: A neuroscience view of the leader–follower duality, Working Paper 1, Centre for Applied Finance and Economics, Birmingham City Business School

Dobelli, R (2013) *The Art of Thinking Clearly*, Spectre, London

Fedor, DB, Davis, WD, Maslyn, JM and Mathieson, K (2001) Performance improvement efforts in response to negative feedback: The role of source power and recipient self-esteem, *Journal of Management*, 27, pp 79–97

Festinger, L (1957) *A Theory of Cognitive Dissonance*, Row, Peterson, & Co, Evanston, IL

Frazzetto, G and Anker, S (2009) Neuroculture, *Nature Reviews Neuroscience*, 10(11), pp 815–21

French, JRP and Raven, BH (1959) The bases of social power, in *Studies of Social Power*, ed D Cartwright, pp 259–69, Institute for Social Research, Ann Arbor

Goetz, T (2011) The feedback loop, *Wired Magazine*, July, pp 126–133, 162

Haidt, J (2006) *The Happiness Hypothesis: Finding modern truth in ancient wisdom*, Basic Books, New York

Harrington, A (1992) At the intersection of knowledge and values: Fragments of a dialogue in Woods Hole, Massachusetts, August 1990, in *So Human a Brain: Knowledge and values in the neuroscience*, ed A Harrington, pp 247–324, Springer Science and Business Media, New York

Heath, C and Heath, D (2010) *Switch: How to change things when change is hard*, Crown Business, New York

Hickson, DJ, Hinings, CR, Lee, CA, Schneck, RS and Pennings, JM (1971) A Strategic contingencies theory of intra-organizational power, *Administrative Science Quarterly*, 16, pp 216–29

Ignatius, A (2013) Influence and leadership. Editorial on the HBR Special issue on influence: How to get it, how to use it, *Harvard Business Review online*, https://hbr.org/2013/07/influence-and-leadership (archived at https://perma.cc/5K4S-354G)

Kahneman, D (2011) *Thinking, Fast and Slow*, Farrar, Straus and Giroux, New York

Klucharev, V, Smidts, A and Fernández, G (2008) Brain mechanisms of persuasion: How 'expert power' modulates memory and attitudes, *Social Cognitive and Affective Neuroscience*, 3(4), pp 353–66

Langer, EJ, Blank, A and Chanowitz, B (1978) The mindlessness of ostensibly thoughtful action: The role of 'placebic' information in interpersonal interaction, *Journal of Personality and Social Psychology*, 36(6), pp 635–42

Leckart, S (2012) The hackathon is on: Pitching and programming the next killer app, *Wired online*, www.wired.com/2012/02/ff_hackathons/all/1 (archived at https://perma.cc/G57S-BMFN)

MacLean, PD (1990) *The Triune Brain in Evolution: Role in paleocerebral functions*, Plenum Press, New York

Mechanic, D (1962) Sources of power of lower participants in complex organizations, *Administrative Science Quarterly*, 7, pp 349–64

Mercier, H and Sperber, D (2017) *The Enigma of Reason: A New theory of human understanding*, Allen Lane, London

Montag, C and Panksepp, J (2017) Primary emotional systems and personality: An evolutionary perspective, *Frontiers in Psychology*, 8, p 464

Newberg, A and Waldman, MR (2012) *Words Can Change Your Brain: 12 conversational strategies to build trust, resolve conflict and increase intimacy*, Hudson Street Press, New York

Patchen, M (1974) The locus and basis of influence on organizational decisions, *Organizational Behavior and Human Performance*, 11, pp 195–221

Pettigrew, A (1972) Information control as a power resource, *Sociology*, 6, pp 187–204

Pribram, KH (2002) Pribram and MacLean in perspective, in *The Evolutionary Neuroethology of Paul MacLean: Convergences and Frontiers*, ed GA Cory and R Gardner, pp 1–8, Greenwood Publishing Group, Westport

Psychogios, A, Antonacopoulou, A, Nyfoudi, M, Blakcori, F and Szamosi, TL (2018) How does feedback matter for the sustainability of organizational routines? Annual Meeting of the Academy of Management, August 2018, Chicago, Conference proceedings – paper 16801

Psychogios, A, Blakcori, F, Szamosi, L and O'Regan, N (2019) From feeding-back to feeding-forward: Managerial feedback as a trigger of change in SMEs, *Journal of Small Business and Enterprise Development*, 26(1), pp 18–42

Renvoise, P and Morin, C (2007) *Neuromarketing: Understanding the 'buy buttons' in your customer's brain*, SalesBrain LLC, San Francisco

Sullivan, J (2015) Born to trust: The brain evolution as a social organism – a conversation with Louis Cozolino, PhD, *BrainWorld*, 6(2), Winter, pp 50–53

Yeung, R (2011) *I is for Influence: The new science of persuasion*, Macmillan, London

SUMMARY OF PILLAR 4: RELATIONS

Leadership lessons

Focus on collective effort

Focus on collaboration rather than on individualistic effort.

Create close friendships

Five recommendations for how to create close friendships:

- nurture a sense of security;
- active listening;
- be vulnerable;
- reciprocal support;
- treat your different friendships differently.

Connections matter

Do not neglect connections with people who can have a dramatic impact on brain functions, and on leadership performance.

Understand how others are thinking

Engage directly with others by creating a comfortable situation, being as clear as possible and listening actively.

Be aware of imitation

Leaders should be aware of imitation when shaping other people's moods, attitudes and behaviour.

Persuade

Persuade by adopting the triune brain approach:

- rationalize the direction;
- emotionalize motivation;
- formulate the environment.

Evaluate the collective brain

Use the threefold diagnostic tool for evaluating the collective brain:

- what people know;
- what people feel;
- what people do.

Be aware of influence

Be aware of the following principles of influence:

- reciprocity;
- commitment and consistency;
- social proof;
- authority;
- liking;
- scarcity;
- unity.

Talk to the brain

There are specific words or phrases that have immediate impact:

'That's not all.'

'Because...'

'I need you to...'

'You have a choice.'

Be aware of feedback

Emphasize:

- informal feedback;
- specific feedback;
- benefits-oriented feedback.

Persuasion stimuli

Six specific stimuli can make leaders more effective in persuasion:

- self-centredness;
- contrast;
- tangibility;
- beginning and end;
- visualization;
- emotionality.

Social influence dimensions:

- similarity vs differentiation;
- familiarity vs novelty;
- collaboration vs competition.

Do not use only one approach or the other. Use both alternatives in all three categories, even simultaneously, to maximize influence.

Concluding remarks

The future of brain science, leadership and the BAL approach

The future of brain science

Ground-breaking discoveries about the brain have just started to surface. What we currently know is probably just the tip of the iceberg. As Gary Marcus and Jeremy Freeman, editors of the book *The Future of the Brain: Essays by the world's leading neuroscientists* (2015) declared, 'There's never been a more exciting moment in neuroscience than now'. We are expecting the next two decades to be seminal in revealing even more intriguing insights and facts about how our neurons fire, interact and impact on our behaviour. We did our best, within the usual constraints, to include in this book the latest findings and discussions about the brain, combining them with classic studies that go back to the middle of the 20th century. We feel that this blend of classic and modern (with an emphasis on the latter), based on extensive research conducted in brain-related literature, together with the tangible results we have observed, gives the BAL approach its edge. We believe that the four pillars will remain as is for a long time to come. Keep an eye out, though, and an open mind, for new findings, new research, new applications of neuroscience and behavioural sciences. Use those new findings to develop it even further yourself! Your brain is plastic and it loves to grow. Use the BAL approach to do so.

From all the new developments in the wider field of neuroscience, the one that will have a profound impact on the way we perceive and deal with the brain is brain-interacting or brain-altering technology. Initial findings, in what has been termed *neurotechnology* (Donoghue, 2015), are staggering, and create huge hopes for people with brain damage. BrainGate (www. braingate2.org (archived at https://perma.cc/L7MF-9S8Q)), Elon Musk-backed Neuralink (www.neuralink.com (archived at https://perma.cc/H38F-RSS5)) and other similar scientific and business initiatives around the world are building devices and computer software that interact directly with neurons in the brain. It seems that not only naturally occurring neuroplasticity can improve our brains, but also electrodes implanted in our heads and sensors on the top of our skulls. Technology will be able to bypass our executive brains. It will reach deeper structures to evoke direct responses. It will acquire clearer information of thoughts and feelings in our brain. And it will help two or more brains to directly interact and communicate. Here is a selection of highly intriguing studies on the topic:

1 *Direct brain-to-brain communication.* A team led by Rajesh PN Rao, Director of the NSF Center for Sensorimotor Neural Engineering and Professor of Computer Science and Engineering at the University of Washington, conducted the first-ever brain-to-brain interface in August 2013 (Rao *et al*, 2014). In the experiment, two people sitting in different locations had to cooperate with each other in order to win a computer game. The first, called the sender, was looking at the screen of the game thinking what move he and his game partner should do, but did not have access to the physical controls. The second, called the receiver, made the moves in the game by using his hand but without having access to the screen to see for himself which move was required. So he depended on the sender. The sender was wearing a brainwaves-reading device (EEG) that picked up his thoughts on making the right moves to win the game. The receiver was also wearing a device, called transcranial magnetic stimulation (TMS), that could make him move his hand involuntarily by stimulating magnetically specific motor centres in his brain. They managed to win the game. The thoughts of the sender moved the hand of the receiver in the right direction to play, according to the simple rules of the video game. In the follow-up to this much-publicized achievement, the team managed to connect successfully three brains, in what they termed BrainNet, to play the classic computer game Tetris (Jiang *et al*, 2019). As in the earlier experiment, all participants were wearing EEG devices, with two of them acting as senders and one as the receiver. Not only was the game played successfully at a rate of 81.25 per cent but, most importantly, the receiver could tell which of the two senders was

most accurate in sending the right brain signals to win the game based on brain input alone! The social implications of this are profound. In the researchers' own words (2019):

> We found that like conventional social networks, BrainNet allows Receivers to learn to trust the Sender who is more reliable, in this case, based solely on the information transmitted directly to their brains. Our results point the way to future brain-to-brain interfaces that enable cooperative problem solving by humans using a 'social network' of connected brains.

2 *Image and speech reconstruction.* In a paper published in 2011, Sinji Nishimoto and associates announced their success in capturing images from people's minds while watching nature films, and reconstructing them again on a separate screen. People in an fMRI scan were watching these films whilst, at the same time, the scientists were recording what was happening in those people's brain areas associated with visual imaging, specifically at the occipito-temporal visual cortex. By using advanced models of visual representation, they managed to reconstruct the dynamic images viewed by participants to a satisfactory degree (Nishimoto *et al*, 2011). Although somewhat fuzzy, these reconstructed images were good enough to create high hopes for improvement in the near future. As Thomas Naselaris, one of the authors of the study, stated more recently: 'The potential to do something like mind reading is going to be available sooner rather than later. It's going to be possible within our lifetimes' (Requarth, 2015). Concerning speech reconstruction, in a Facebook-funded study, Moses and associates (2019) placed electrodes directly on the brains of people about to undergo brain surgery. They then asked those people to read aloud a series of questions and as well as a list of possible answers to each question. By analysing the electromagnetic data generated in the brain by the movements of the larynx, lips, tongue and jaw they could predict which question, at 76 per cent accuracy, and which answers, at 61 per cent accuracy, were read aloud by participants just by looking at brain data. This means that, in the future, brain data could reveal what someone says without the need to actually hear the person speaking at all!

3 *Sharing thoughts.* A study published in 2014 demonstrated that two connected brains can share a thought through non-invasive neurotechnologies (Grau *et al*, 2014). A person sitting in Kerala in India (the emitter) was asked to imagine that he is moving either his hand or

foot. The brain–computer interface device he was wearing on his head captured the signal in his brain, generated by his thought, and translated it into a simple binary code (0 and 1) that represented each imagined action. Then the researchers assigned a word to each imagined action, the Italian word *ciao* and the Spanish word *hola*, depending on whether he thought of moving his hand or his foot. The emitter knew this so he could decide which word it would be when he made a move. The generated binary codes were then emailed to France, where another person (the receiver) was sitting in Strasburg with his eyes closed and wearing another device on his head. The device translated the binary code it received into flashes of light that the receiver saw in his mind's eye. Knowing in advance what the flashes could mean, he could say the words *ciao* or *hola*, thus replicating directly what the emitter had decided to send. In essence, the two brains communicated directly with each other over the internet. The emitter intended to say either *ciao* or *hola* and the receiver knew which one. Giulio Ruffini, one of the authors of the study, said very openly that (Eveleth, 2015):

> you can look at this experiment in two ways. On the one hand it's quite technical and a very humble proof of concept. On the other hand, this was the first time it was done, so it was a little bit of a historical moment I suppose, and it was pretty exciting.

With such leaps in brain understanding and interaction, the renowned physicist and futurologist Michio Kaku rightfully included in his 2015 book *The Future of the Mind: The scientific quest to understand, enhance, and empower the mind* topics as controversial and mystical as telepathy, telekinesis, mind control, silicon consciousness and mind beyond matter. In his own words, from the opening lines of the book:

> The two greatest mysteries in all of nature are the mind and the universe... If you want to appreciate the majesty of the universe, just turn your gaze to the heavens at night, ablaze with billions of stars... To witness the mystery of our mind, all we have to do is stare at ourselves in the mirror and wonder, What lurks behind our eyes?... But [until recently] the basic tools of neuroscience did not provide a systematic way of analyzing the brain.

Now, it does. Brain-reading technology is accelerating fast with both businesses and governments strengthening their efforts to get ahead in brain–computer interface (BCI) (Rogers, 2019). At the same time, though, we definitely need to update protection regulations to catch up with these developments (Duncan, 2019).

What do all these advances mean for brain-based leadership? Although it is difficult to say, we can envision a few things. Firstly, leaders will take brain training methods and initiatives, like the BAL approach, more seriously – both for themselves and for their organizations. This also means, hopefully, that brain-friendly approaches will be introduced early on in education in order for children and teenagers to use brain-enhancing approaches to deal more effectively with the challenges they face at school and in life.

Secondly, leaders and managers will be using mobile applications, computer programs and various devices to help them get neurofeedback (how their neurons behave triggered by controlled stimuli) for many of their daily tasks. Already feasible examples include:

- Recruiting and developing personnel involving facial recognition and electroencephalogram (EEG) tests. Applied neuroscience companies like Trizma Neuro, where one of us, Dr Nikolaos Dimitriadis, is the co-founder and CEO, are creating proprietary skills-assessment methodology using neutotech. All it takes is a five minutes session in front of a screen, wearing an EEG on your head and having your biometric data recorded through various devices, to neuro-assess your empathy, resilience and creativity! Other companies around the world are using neuroscience-based games to assess similar competencies for both recruitment and people development.

- Detecting brainwaves of people at the office in order to enable them to get faster into their flow state and thus be more productive and fulfilled.

- Mapping emotions through voice recognition and other methods during meetings to help companies create a more collaborative and exciting environment for their employees.

- Collecting physiological/biometric data, mainly though wearables, to detect burnout, anxiety and other conditions that threaten the well-being of people in organizations.

At the same time, advanced data analytics from all possible digital sources will push the boundaries in discovering deeply rooted behavioural patterns at work.

Thirdly, further in the future, we see the direct implementation of neuro-technology and brain-to-brain communications in leadership, probably starting from global corporations, governments and even education. Of course, all these are predictions and, as such, they need to be considered with caution. More so, since many of them will not become reality if the two important issues of privacy and ethics are not adequately resolved.

Of course, applied neuroscience and neurotechnology are like any other human tool: they can be used with good or bad intentions. For example, they can be used for promoting egoistic goals and for manipulation, or for creating shared and collective benefits. They can also be used as a means of conflict resolution or conflict expansion, and even for war. A special report in the magazine *Foreign Policy* revealed that neuroweapons are already being made and that considerable amounts of money are been poured from defence agencies and companies into integrating neurotechnology with human and infrastructural systems (Requarth, 2015). 'Welcome to the neurowars', the front page of the September–October 2015 issue of *Foreign Policy* declared, so, unfortunately, we might not be so far from such an undesirable future.

BRAIN STRUCTURE, COMPANY STRUCTURE

Organizational structures will not be immune to neuroscience. The way we organize jobs, tasks, authority levels and communication lines within organizations will follow the way our brains are organized. Internal brain structures will be mirrored in team, departmental and organizational design and charts. Does it sound too far fetched? It is already happening. Renowned leadership guru John P Kotter explained in his seminal *Harvard Business Review* article 'Accelerate' (2012), and expanded in his book *XLR8: Building strategic agility for a faster-moving world* (2014) his finding of a dual operating system within corporations. This dual system consists of the traditional, pyramid-like, formal and inflexible hierarchy, with more informal, spontaneous, flexible, startup-like and project-driven networks. The first system needs to exist to ensure continuity of delivery and stability while the second is vital for boosting innovation, creative disruption and future competitiveness. Companies need both as our brains need both: rationality and emotions need to work together, in harmony. If not, brains get dysfunctional and companies get outpaced by leaner and more creative startups.

Marketing experts Marc de Swaan Arons, Frank van den Driest and Keith Weed published their article 'The ultimate marketing machine' in the *Harvard Business Review*, in an issue dedicated to 'Think, feel, do: The new basics of marketing'. In that article, the authors described the optimum marketing function structure as one mirroring key brain functions (Arons *et al*, 2014). They suggested that in order for marketing to work better, 'think' tasks (including research and analytics), 'feel' tasks (including customer engagement management and media relations) and 'do' tasks (including content creation and production) need to be handled by different teams, each one staffed with specialists in their respective brain function. Mildly put, this is a revolution in the making. Leaders need to urgently become neuroscience-ready and lead this revolution into new realms. Those who do will be the winners.

Homo relationalis: Brain-adaptive relational leadership

Traditional leadership theories attempted to approach and understand leadership as an individual feature that consists of many cognitive aspects. The focus was on the individual that acts as a leader and their traits, behaviours and styles. We believe that these approaches are better left behind. The future of leadership is relational and unifies three core elements, namely the leader(s), the follower(s), and the context. You cannot understand the wholeness of the phenomenon called leadership if you focus on one or two of the above elements. Leadership is a dynamic process that takes place in the relationship and interaction between the leader and the follower within a particular context. If you miss one of the three interconnected elements you miss the actual sense of leadership. The leader (and the follower) cannot be seen independently of the context in which they participate, and they cannot be seen independently of the people they relate with. Leadership occurs within the process of relating with each other, aiming to do things in a non-static, dynamic and continuously evolving way. This relational approach to leadership is not a new one; for more than 15 years now there has been a school of thought arguing that we need to understand the nature of leadership as a relational process (Dihn *et al*, 2014; Uhl-Bien, 2005, 2006). Relational leadership can be understood as a social construction process within complex collective entities (organizations) and through connections and interdependencies of their members (Psychogios and Garev, 2012; Bradbury and Lichtenstein, 2000).

In another work of ours we proposed the term *Homo relationalis* to show that it is the relational aspects (interconnectivity, interrelatedness and interaction) between us that need to be factored in if we want to explain and understand further leadership process (Dimitriadis and Psychogios, 2020). However, there is a missing link regarding the ways that leadership is constructed as a social process. We suggest that this gap can be covered if we also consider different approaches to understanding social connection that can be found in brain science. Based on the relational aspect of leadership, true leadership actors are caring for each other using their brains as a tool for developing in-group collaboration, interdependence and trust. The main reason is that relational leadership behaviours are closer to the real nature of human brains that are evolved to be social. Based on a more complex cortex, humans have the ability to reciprocate, collaborate, empathize, trust, form intelligent analysis of social situations, but also deceive and fight more cleverly than other species (Dunbar, 1998). These human brain natural capabilities should be the key relational bonds in a continuous process of interaction among leaders and followers.

In other words, the future of leadership is relational and brain-adjusted. If we are to embrace the full scope of our interdependent social brains, we need to emphasize brain adjusted relational leadership (Dimitriadis and Psychogios, 2020). Brain-adaptive relational leadership is a dynamic, co-created type of leadership based on relational ties of all leadership actors (leaders and followers). It is a type of leadership that shows a better understanding of the inner workings and, especially, the purpose of key brain functions. A type of leadership that will bring about trust and collaboration within organizations, as well as controlled competition when needed, and that, ultimately, will unleash the true power of the *Homo relationalis*. We think that our BAL approach drives the future of leadership.

In this respect, we see the fourth pillar of our BAL model, relations, as more of a horizontal line that forms the foundation for all other three pillars.

THE NEW COLLABORATIVE PARADIGM
From Homo economicus to Homo relationalis

If you ask someone which species shows the highest level of collaboration between their members, the typical answer that you will receive is: bees and ants. Bees' and ants' ability to collaborate within their communities harmoniously and relentlessly attracts the attention of the public. But is this really the case? If cooperation is genetically predefined, are humans a collaborative or a competitive species? The admiration of bees and ants as the ultimate cooperative machines is based on a fundamental misconception. Actually, those species do not have the decision power to choose collaboration or competition, but are directed by chemicals to collaboration (Gamble *et al*, 2014). In this chemically induced tyranny of co-working, those species are born to cooperate with specific members of their community and for specific reasons. In an analogy to human societies, such organizations would look more like totalitarian systems. Actually, not giving the answer 'humans' to this question reveals the damage done by the *Homo economicus* mindset that focuses exclusively on competitive struggle, self-interest and isolative individuality.

Homo economicus is an individualistic conception of humanity, void of any social dimension, which considers as natural law that the self-interest of one person is the interest of all people, leading to an ultra-egocentric model of decision-making and behaviour. The *Homo economicus* concept can also be found in traditional leadership studies, and it is consistent with a positivistic epistemology and a Cartesian dogma of a clear distinction between mind and nature (Bradbury and Lichtenstein, 2000). In other words, this view drives us away from real nature that in fact is much more collaborative than we might imagine. Looking at the evolution of *Homo sapiens sapiens* in comparison with other species, and studying brain size in relation to group size, has led to the social brain theory. This theory suggests that socializing, collaborating and co-existing in communities depends on brain size, especially frontal lobe cortical areas (Dunbar, 1998). In other words, humans developed larger brains in order to be able to form strong social relations and increasing their chances to survive by being part of a social group. In this vein, the *Homo economicus* model of humanity is fading away: new models such as the *Homo reciprocans*, the *Homo sociologicus* and the *Homo socioeconomicus* have emerged as an effort to understand better the complex interrelations between people (O'Boyle, 2007). We suggest the *Homo relationalis* in order to emphasize the need to relate in our daily interactions that enhance collaboration and express what we call brain adjusted relational leadership.

The future of the brain-adaptive leadership approach

The very fact you are reading this sentence means that our application of the BAL model in writing the book was successful. Whether the book itself is successful depends on the results you'll get by applying the BAL model. You succeed, we succeed.

But this was not the first time we have applied the BAL model. Our involvement with neuroscience, psychology, sociology, anthropology, behavioural economics and, of course, with the sciences of management and marketing, has inevitably calibrated our brains to apply many of the principles of the book almost automatically. For the last 10 years both of us have been engaged in multiple roles in our professional lives: lecturers, speakers, trainers, researchers, mentors, coaches, consultants, managers and entrepreneurs. These roles helped us generate multiple experiences with the BAL model, both directly and through the people we taught, advised and coached. Thus, we have not included in the book any insight or recommendation that we have not tried ourselves and that did not produce tangible results. It is this deep-rooted, experiential and shared belief we have in the BAL approach that gave us the impetus to write the book and make it accessible to more people around the world. Our promise is this: by applying the BAL principles your brain will become your strongest ally in leadership.

All pillars of the BAL approach will almost certainly improve and expand in the future. Let's see briefly how a few of the issues discussed earlier will help (with some already happening):

Pillar 1. Thinking will become smarter. And it will focus on the things that it can do rather than what we thought it was doing. We will spot biases, false patterns and ego depletion faster than ever due to self-assessment applications and better analysis of data from the world around us as well as from our own body and brain. We will ask better questions, become more resilient, place smarter strategic bets and understand sooner whether we are pursuing our purpose as an organization, since communication between people will become more brain-friendly.

Pillar 2. Emotions will finally be appreciated as the basis of all motivation and decision-making for action. Consequently, feelings will become more meaningful. Our capacity for emotional intelligence will increase due to feedback technology that will allow us to understand better what we and others feel at work. Also, emotional styles and moods will be tracked and managed better through data analytics and emotion-detecting interfaces. We will not obsess only about positive emotions but we will recognize the need for a holistic emotional approach that utilizes negative emotions as well.

Pillar 3. Brain automations will work more to our advantage. Neurodesign will help us create better spaces, both online and offline. Those spaces will prime our brains for better cooperation, more creativity and higher productivity. We will also be able to track habits more easily through apps and sensors in mobile and other devices.

Pillar 4. Relations will become stronger and more motivating. Social media are already revolutionizing the way people connect, work and live, inside and outside organizations, but many times with negative effects. Although building useful bridges will be easier than ever, we will also be able to nurture and evaluate those bridges more efficiently. Persuasion will be digitally aided through advanced analytics and even computer generated information and suggestions. This will be largely based on what is called predictive neuroscience, which is evaluating, modelling and predicting neural behaviour in our brain (Markram, 2013). At the same time, internal brain structures will form the basis for corporate re-structuring approaches worldwide. Neurotechnology will optimize communication between parts in these new structures in order to become super-efficient and super-effective, as our brain already is.

Complexity asks for great leaders. Great leaders in any social or economic activity and in any company or institution. Great leaders from any age group and from any place on earth. Complexity also requires phronetic leaders. Since past knowledge is not enough to deal with the future complex problems, we need to cultivate more of our practical, day-to-day judgements in taking decisions and in behaving in a social context. We hope that the BAL approach will be a significant step towards strengthening our 'phronesis' and therefore our individual and collective leadership potential, and unleashing it into the world. Go on – unleash it!

Keep in mind

... all the 'Keep in mind' sections in this book. We strongly recommend that you revisit them often to refresh your memory and to boost your brain leadership into new realms!

References

Arons, MDS, van den Driest, F and Weed, K (2014) The ultimate marketing machine, *Harvard Business Review*, July–August, **92**(7), pp 54–63

Bradbury, H and Lichtenstein, B (2000) Relationality in organizational research: Exploring the 'space between', *Organization Science*, **11**(5), pp 551–64

Dihn, J, Lord, RG, Gardner, W, Meuser JD, Liden, RC and Hu, J (2014) Leadership theory and research in the new millennium: Current theoretical trends and changing perspectives, *The Leadership Quarterly*, **25**, pp 36–62

Dimitriadis, N and Psychogios, A (2020) Social brain-constructed relational leadership: A neuroscience view of the leader–follower duality, CAFÉ Working Papers 1, Centre for Applied Finance and Economics, Birmingham City Business School

Donoghue, J (2015) Neurotechnology, in *The Future of the Brain: Essays by the world's leading neuroscientists*, ed G Marcus and J Freeman, pp 219–33, Princeton University Press, Princeton

Dunbar, RIM (1998) The social brain hypothesis, *Evolutionary Anthropology: Issues, news, and reviews*, **6**(5), pp 178–90

Duncan, A (2019) Mind-reading technology is closer than you think, www.fastcompany.com/90388440/mind-reading-technology-is-closer-than-you-think (archived at https://perma.cc/CF9U-DYYX)

Eveleth, R (2015) I emailed a message between two brains, www.bbc.com/future/story/20150106-the-first-brain-to-brain-emails (archived at https://perma.cc/NB5F-PFU2)

Gamble, C, Gowlett, J and Dunbar, R (2014) *Thinking Big: How the evolution of social life shaped the human mind*, Thames and Hudson, London

Grau, C, Ginhoux, R, Riera, A, Nguyen, TL, Chauvat, H, Berg, M, Amengual, JL, Pascual-Leone, A and Ruffini, R (2014) Conscious brain-to-brain communication in humans using non-invasive technologies, *PLOS ONE*, **9**(8), e105225, doi 101371/journalpone0105225

Jiang, L, Stocco, A, Losey, DM, Abernethy, JA, Prat, CS and Rao, RP (2019) Brainnet: A multi-person brain-to-brain interface for direct collaboration between brains, *Scientific Reports*, **9**(1), pp 1–11

Kaku, M (2015) *The Future of the Mind: The scientific quest to understand, enhance, and empower the mind*, Anchor Books, New York

Kotter, JP (2012) Accelerate! *Harvard Business Review*, November

Kotter, JP (2014) *XLR8: Building strategic agility for a faster-moving world*, Harvard Business Review Press, Boston

Marcus, G and Freeman, J (2015) Preface, in *The Future of the Brain: Essays by the world's leading neuroscientists*, ed G Marcus and J Freeman, pp xi–xiii, Princeton University Press, Princeton

Markram, H (2013) Seven challenges for neuroscience, *Functional Neurology*, 28(3), pp 145–51

Moses, DA, Leonard, MK, Makin, JG and Chang, EF (2019) Real-time decoding of question-and-answer speech dialogue using human cortical activity, *Nature Communications*, 10(1), pp 1–14

Nishimoto, S, Vu, AT, Naselaris, T, Benjamini, Y, Yu, B and Gallant, JL (2011) Reconstructing visual experiences from brain activity evoked by natural movies, *Current Biology*, 21(19), pp 1641–46

O'Boyle, EJ (2007) Requiem for Homo Economicus, *Journal of Markets and Morality*, 10(2), pp 321–37

Psychogios, A and Garev, S (2012) Understanding complexity leadership behaviour in SMEs: Lessons from a turbulent business environment, *Emergence: Complexity and organization*, 14(3), pp 1–22

Rao, RPN, Stocco, A, Bryan, M, Sarma, D, Youngquist, TM and Wu, J (2014) A direct brain-to-brain interface in humans, *PLOS ONE*, 9(11), e111332, doi 101371/journalpone0111332

Requarth, T (2015) This is your brain. This is your brain as a weapon, Foreign Policy, https://foreignpolicy.com/2015/09/14/this-is-your-brain-this-is-yourbrain-as-a-weapon-darpa-dual-use-neuroscience/ (archived at https://perma.cc/AY6E-ZN59)

Rogers, S (2019) Brain–computer technology is accelerating: Will we soon be typing with our minds? www.forbes.com/sites/solrogers/2019/09/25/brain-computer-technology-is-accelerating-will-we-soon-be-typing-with-our-minds/#4b78fadd483c (archived at https://perma.cc/PX3Z-BAUQ)

Uhl-Bien, M (2005) Implicit theories of relationships in the workplace, in *Implicit Leadership Theories: Essays and explorations*, ed B Schyns and JR Meindl, pp 103–33, Information Age Publishing, Greenwich, CT

Uhl-Bien, M (2006) Relational leadership theory: Exploring the social processes of leadership and organizing, *The Leadership Quarterly*, 17(6), pp 654–76

EPILOGUE

The BAL approach has become integral to the way that both of us experience our professional and personal lives. Writing this book was no exception. It required us putting into practice all four pillars of the BAL approach carefully, consistently and creatively. This is how.

On thinking, we had to use our willpower carefully since we both worked full time while writing it. We had to avoid as much as possible the burnout effect, various cognitive biases, and misleading pattern recognition, especially while deciding which relevant literature to use and which suggestions and examples. We had to constantly ask the right questions for the development of the book and remind ourselves of the higher purpose for writing it. We had to be creative in solving various challenges, get into flow as often as possible (for fast progress) and overstretch our memory to make sure that we didn't leave out anything we deemed important. Enhancing our growth mindset was absolutely necessary!

On emotions, the majority of emotions experienced whilst writing this book ranged, thankfully, from positive to extremely positive. So, the dominant mood was that of the *rocket*. However, any negative emotions that arose during any mishaps were also put to good use: they made us more determined and more resilient towards achieving our aim. All these emotions were what kept us going (as emotions do). The dominant core emotion was *excitement*, or *seeking*, and the emotional equation we experienced the most was *curiosity* (= wonder + awe).

On brain automations, we took advantage of priming by making sure that our desks were always full of materials related to the book. We made daily online meetings between us a routine, wherever in the world we happened to be. We impulsively bought any book or magazine on the market (usually in airport bookstores throughout Europe and Asia) that mentioned neuroscience or the brain, as well as reading the latest special reports on the subject in various online sources. These are also habits we've shared for the best part of 10 years.

On relations, we did our best to stay closely connected to a number of people, including each other, our publisher's wonderful team, professionals and scientists with similar interests around the globe and, of course, our students, colleagues and clients. Weak links proved as important as strong ones in terms of new ideas, problem solving and reaching out to the world.

Most importantly, we never stopped trying to positively influence a variety of people, such as (again) each other, our business and life partners, and our support team at our publisher in order to finish the book in the best way possible. At the same time, by mentioning the book to people we met in class, companies and events, we tried to sway them positively and create anticipation.

Thank you for reading this book and, hopefully, for sharing our view on this new, brain-based, era of leadership. There are now two clear paths ahead of you: either you continue having the same thoughts, emotions, habits and relations as you did before reading it *or* you apply the four pillars and you become an even better version of your leadership self. The choice is yours. We really hope that your brain has already taken the right one!

INDEX